T0135454

Supporting the Understanding of Team Dynamics in Agile Software Development Through Computer-Aided Sprint Feedback

Fabian Kortum

Logos Verlag Berlin

Bibliografische Information der Deutschen Nationalbibliothek

Die Deutsche Nationalbibliothek verzeichnet diese Publikation in der
Deutschen Nationalbibliografie; detaillierte bibliografische Daten sind im
Internet über http://dnb.d-nb.de abrufbar.

ISBN 978-3-8325-5438-5

Logos Verlag Berlin GmbH
Georg-Knorr-Str. 4, Gebäude 10
D-12681 Berlin

Tel.:   +49 (0)30 / 42 85 10 90
Fax: +49 (0)30 / 42 85 10 92
http://www.logos-verlag.de

"Tell me how you will measure me and I will tell you how I will behave."

- E. M. Goldratt -

Abstract

The complexity of software projects and inherent customer demands is becoming increasingly challenging for developers and managers. Human factors in the development process are growing in importance. Consequently, understanding team dynamics is a central aspect of steady development planning and execution.

Despite the many available management systems and development tools that are being continuously improved to support teams and managers with practical process information, the equally crucial sociological aspects have typically been addressed insufficiently or not at all. In people-focused agile software processes, a first socio-technical understanding can also be promoted by sharing positive and negative development experiences during specific team meetings (e.g., sprint Retrospectives). Nevertheless, there is still a lack of systematically recorded and processed socio-technical information in software projects, making it difficult for subsequent reviews by teams and managers to characterize and understand the sometimes volatile and complex team dynamics during the process.

This thesis strives to support teams and managers in understanding and improving awareness of the team dynamics that occur in their agile software projects by introducing computer-aided sprint feedback. The concept builds on four information assets: (1) socio-technical data monitoring, (2) descriptive sprint feedback, (3) predictive sprint feedback, and (4) exploratory sprint planning. These assets unify interdisciplinary fundamentals, practical methods from software engineering, data science, organizational and social psychology. Using a design science research process for information systems, observations in several conducted studies (32 in academic project environments and three in industry) resulted in the foundations and methods for a practical feedback concept on the socio-technical aspects in sprint, prototypically realized for Jira.

A practical evaluation involved two industry projects in an action research methodology that helped improve the concept's usability and utility through practitioner reflections. The collaboration between industry and research resolved practical issues that did not arise during the design science process. Several beneficial outcomes based on the provided sprint feedback are reported and described in this study (e.g., the effect of team structures on development performance). Moreover, the reflections underscored the practical relevance of systematic feedback and the need to better understand human factors in the software development process.

Keywords: Team Dynamics, Agile Software Development, Knowledge Support

Zusammenfassung

Steigende Kundenanforderungen und immer komplexere Softwareprojekte stellen zunehmende Herausforderungen für Manager und Entwicklerteams dar, in denen die menschlichen Aspekte im Entwicklungsprozess an Bedeutung gewinnen. Für die Projektplanung und -Durchführung sind Teamdynamiken wichtig zu kennen.

Projektmanagementsysteme und Entwicklungstools unterstützen Teams und Manager weitreichend mit Prozessinformationen, dabei werden die ebenso relevanten soziologischen Aspekte in der Regel jedoch nur unzureichend abgedeckt. Agile Entwicklungsprozesse fördern das soziotechnische Grundverständnis durch einen Regelaustausch positiver und negativer Erfahrungen in Teammeetings (z.B. Retrospektiven). Eine systematische Erfassung und Verarbeitung soziotechnischer Informationen zur Interpretation von Teamdynamiken im Entwicklungsprozess (z.B. in Reviews durch Teams oder Management) wird hingegen kaum praktiziert.

Ziel dieser Arbeit ist es, Teams und Führungskräfte dabei zu unterstützen, ihr Bewusstsein und ihr Verständnis für Verhaltensdynamiken in laufenden Projekten durch computergestütztes Sprint-Feedback zu verbessern. Das entwickelte Feedbackkonzept basiert auf vier Informationssäulen: (1) soziotechnisches Datenmonitoring, (2) deskriptives Feedback, (3) prädiktives Feedback und (4) explorative Planungsunterstützung. Das Feedback stützt sich auf theoretische Grundlagen und praktische Methoden aus dem Software Engineering, den Datenwissenschaften, der Arbeits-, Organisations- und Sozialpsychologie. Design Science wurde verwendet, um mithilfe von Beobachtungsstudien in insgesamt 32 studentischen Softwareprojekten und drei Fallstudien aus der Industrie, ein für die Praxis anwendbares Feedbackverfahren zu soziotechnischen Abhängigkeiten in Sprints zu konzeptionieren und als prototypische Erweiterung für Jira umzusetzen.

Action Research wurde für die praktische Bewertung der Anwendbarkeit und Nützlichkeit von computergestütztem Feedback in zwei industriellen Fallstudien angewendet. Durch die Zusammenarbeit zwischen Industrie und Forschung konnten neue praktische Probleme erkannt und gelöst werden, die mit dem Design Science Ansatz nicht erkennbar waren. In beiden Fallstudien wurden mehrere positive Erkenntnisse berichtet aufgrund des Sprint Feedbacks (z. B. der Einfluss von Teamstrukturen auf die Entwicklungsleistung). Die Fallstudien unterstrichen zudem die Praxisrelevanz für ein verbessertes Verständnis zu menschlichen Faktoren in der Softwareentwicklung sowie der damit einhergehenden Notwendigkeit für systematisch bereitgestelltes Sprint-Feedback im laufenden Projekt.

Schlagworte: Teamdynamiken, Agile Softwareentwicklung, Wissensförderung

Contents

List of Figures

List of Tables

Abbreviations

2FS	2-Features Selection
AFS	Automated Feature Selection
ANN	Artifical Neural Network
AMS	Automated Model Selection
APD	Agile Project Dynamics
AQ	Assessment Question
AR	Action Research
ARC	Action Research Cycle
ARFF	Attribute-Relation File Format
AS	Abstraction Sheet
ASD	Agile Software Development
AutoML	Automated Machine Learning
DIKW	Data Information Knowledge Wisdom
DSR	Design Science Research
DT	Decision Tree
GQM	Goal Question Metric
IPA	Interaction Process Analysis
ISR	Information System Research
KDD	Knowledge Discovery in Databases
KMS	Knowledge Management System
KNN	K-Nearest Neighbors
LOOCV	Leave-one-Out Cross-Validation
LR	Linear Regression
MI	Mutual Information
MIC	Maximal Information Coefficient
ML	Machine Learning
MLP	Multilayer Perceptron
MLR	Multiple Linear Regression
MSE	Mean Squared Error
MTBF	Mean Time Between Failures
PANAS	Positive and Negative Affect Schedule
RBF	Radial Basis Function
RMSE	Root Mean Square Error
SD	System Dynamics
SNA	Social Network Analysis
SVM	Support Vector Machine
SVR	Support Vector Regression

Chapter 1

Introduction

This doctoral thesis introduces a computer-aided sprint feedback concept to support an understanding of socio-technical dependencies and behavior patterns in teams during agile development processes. This concept has several interdisciplinary foundations and includes practical methods from data science, software engineering, and organizational and social psychology. It comprises an adaptive, holistic process chain for the systematic capture, processing, and characterization of team dynamics in software development projects. The work addresses three research questions concerning the practical relevance, utilization, and utility of complementary team feedback on socio-technical aspects of agile software projects.

1.1 Human Factors in Software Projects

The role of human factors in software development has been investigated over the decade and several systematic literature reviews have shown that studies of this topic have intensified in recent years [16, 111, 198]. Human nature is based on psychological factors (e.g., emotions dictate behaviors), which, in combination with social activities, can impact the success of a development team [111]. In the software engineering domain, broad and diverse research has been conducted on the human and social effects of these factors on software development teams and processes [122, 205, 207, 212]. However, much of the software engineering research conducted in the last decade has focused more on technical problems than on human aspects [111]. Thus, "Failure to include human factors may explain some of the dissatisfaction with conventional information systems development methodologies; they do not address real organizations" [11]. This underscores the need for cooperation between practitioners and researchers to address the software engineering challenges stemming from human factors and improve our understanding of the inherent effects of these factors during development processes [18, 154, 198]. In the software engineering domain, the term *team dynamic* describes interaction patterns among project members that can determine the performance of teams [62].

Human factors, social activities, and development behaviors all contribute to team dynamics, which can influence a team's progress and the achievement of project goals [47, 193]. Solid planning, previous experience, knowledge, and awareness of team dynamics can reduce the likelihood of dysfunctional development situations. A team's size and its fundamental communication structure can affect the team atmosphere and development performance in a project (estimated versus completed story points) [4]. Understanding the team dynamics in ongoing software projects means learning from past interactions and deriving knowledge that enables adjustments for future development activities.

It is particularly important to understand human factors and social interactions, not only in plan-driven processes but also, and indeed especially, in agile software development. Agile methods, such as Scrum [209], Lean [189], or Extreme Programming [23], are people-focused and based on small collaborative teams with positive motivation and continuous striving to improve work habits for better organizational and development performances over time [238, 239]. They are more adaptive than plan-driven processes and allow software teams to react faster to customer demands and change requests within short development iterations [145]. However, the increased development agility of teams also amplifies the risk of volatile human behavior, which can lead to information gaps, loner attitudes, sentiment changes, demotivation, and performance fluctuations [25, 236, 238].

In this context, it is no surprise that agile methods, allow team members to routinely share their positive and negative development experiences through cyclic sprint meetings (e.g., in sprint retrospectives or planning) [164]. Accurate sprint planning is crucial for agile teams to achieve steady development performance in short iterations [164]. However, feature re-prioritization or customer change requests are always a possibility and can derail even the most careful plans. Sprint planning is an experience-based activity that builds on retrospective knowledge gained from earlier development performances, covering both positive and negative situations [50, 59, 228]. Nevertheless, human factors are often insufficiently considered in sprint planning [145]. People have different personalities and skills, which, combined with other environmental factors, can influence performances [47]. Therefore, it is vital for teams and management to learn about how these influences have played out in the past to improve dysfunctional habits (e.g., by holistically reflecting on socio-technical strengths and weaknesses in sprints).

Early recognition and understanding of team dynamics in volatile software projects is challenging, primarily due to the often short development iterations and a lack of psychological acumen, analytical expertise, or practicable methods to enable adaptive socio-technical sprint characterization in existing processes [122, 138]. These problems limit the holistic formation of a shared knowledge base covering information on team dynamics that teams can access. Computer-aided sprint feedback can support understanding of the sometimes even complex dynamics in agile teams with the help of systematically captured and processed socio-technical data through a project. Moreover, visualizations can enhance cognitive perception and awareness as the foundation for improvement opportunities [153].

1.2 Research Motivation and Questions

The aim of this work is to enable faster team feedback on the socio-technical aspects of sprints using computer-aided data capture and processing methods. The feedback support is intended to create an opportunity to sharpen awareness and broaden understanding of team dynamics and behavioral patterns resulting from systematically observed socio-technical dependencies in agile software projects. This work investigates the practical relevance, utilization, and utility of computer-aided team feedback on socio-technical aspects during the development process.

1.2.1 Team Understanding as Motivation

The central aim of this work is to support awareness and understanding behavioral patterns in teams with the help of systematic feedback on socio-technical dependencies during the development process. Data acquisition and analysis of team dynamics in software projects can be a sophisticated and time-consuming process that requires technological support [62, 137]. However, understanding behavior patterns during projects can enable a transition from explicit to tacit knowledge in teams, as shown by the data, information, knowledge, wisdom (DIKW) pyramid in Figure 1.1.

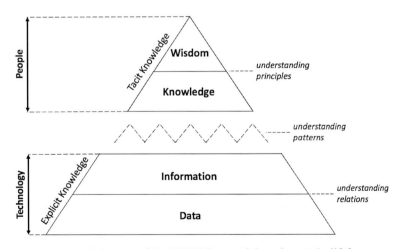

Figure 1.1: Adaption of the DIKW Pyramid, based on Ackoff [6]

Data and information can be seen as explicit knowledge that is accessible using information technologies (e.g., project management systems). By contrast, team knowledge and intellectual wisdom concerning human factors can be classed as tacit knowledge, which anchors in people's minds and includes perceptions, opinions, sentiments, personalities, and experiences [6, 200, 204].

Understanding supports various knowledge transition states (e.g., understanding the relationship between mood changes and development performance in teams enables an identification of patterns over time that must be able to be understood by the human mind to promote implicit knowledge and awareness) [6]. Intellectual wisdom is needed for decision-making and devising improvement actions based on understanding the principles (i.e., the information pattern meaning corresponding to individual experiences and knowledge).

However, the transformation of socio-technical information into knowledge requires an understanding of team behavior patterns (e.g., development performance, communication behavior, mood changes, and team structures). In this work, this understanding is supported by an information technology concept based on computer-aided sprint feedback [6, 47, 50, 139, 228]. Although new information technologies are available that facilitate knowledge acquisition and decision support (e.g., data-driven methods and models for deriving progress tendencies or bug estimations), only a few concern the socio-technical aspects of agile development teams. Moreover, in practice, capturing and interpreting team dynamics is not a trivial activity and often involves adaptions or changes to standard process routines and practices [62, 121, 122].

To be useful in practice (e.g., in sprint retrospectives), computer-aided feedback support on team behavior requires sociological (subjective) and process (objective) data to enable a holistic characterization of team dynamics [59, 134, 139]. Extended feedback mechanisms can strengthen learning in agile teams [61].

This work endeavors to support

a) the analysis of socio-technical data (explicit knowledge) from team reflections (transformed tacit knowledge) to identify behavior patterns.

b) an understanding of team dynamics in agile projects through computer-aided sprint feedback that considers both explicit and tacit knowledge.

Several practical usability and utility challenges arise in connection with these goals. Data observation and processing to obtain useful socio-technical information must be simple and must not entail significant extra efforts that could comprise existing workflows and processes [232]. There must be a clear motivation in teams to contribute of socio-technical data on experiences and perceptions. Furthermore, practicability of the sprint feedback should reflect the benefits in teams, thus enhancing the acceptance [239]. It is highly desirable to systematically support the understanding of team dynamics in agile software projects and promote team knowledge concerning socio-technical aspects.

1.2.2   Research Questions

The challenges and motivation for understanding team dynamics in agile software projects dictate this work's thematic focus. The supplementary support of intra-team knowledge and awareness of socio-technical dependencies in sprints is the central objective of this work. In volatile agile development iterations, fast feedback and knowledge sharing based on cyclic team reflections are particularly crucial information assets for understanding development performance [44, 233].

Sprint retrospectives support open discussions in teams (e.g., about achieved performances) and are useful for identifying whether improvement actions have resulted in the expected changes [59]. Therefore, functional communication structures, information flows, and a positive atmosphere in team meetings are essential for software project success [115, 205, 207]. Sprint feedback from stakeholders and project leaders supplements teams' internal perceptions and knowledge by introducing external perspectives. The latter can refer, for example, to satisfaction or disappointment with stages of the project or to software- or team-related problems, such as a lack of progress or transparency [59, 239].

The following research questions were defined to investigate the practical relevance, utilization, and utility of computer-aided team feedback for understanding team dynamics during the development process.

Research Question 1

How relevant is computer-aided sprint feedback on socio-technical aspects for agile development teams?

The first research question concerns the relevance of supplementary feedback for supporting a team's information needs and understanding of the socio-technical aspects of sprints. In addition, the question aims to identify frequent problems in agile development teams and what information can best support teams in overcoming these problems, based on researchers' and practitioners' experiences. The findings form the foundation for a computer-aided feedback concept to support a practical understanding of team dynamics during the development process.

Research Question 2

How do agile teams utilize computer-aided sprint feedback on socio-technical aspects during the development process?

The second research question focuses on the utilization of computer-aided sprint feedback on socio-technical aspects. The main focus is on determining how agile development teams acknowledge and use complementary sprint feedback on socio-technical elements along with or instead of other information resources during the development process. The question also concerns the feedback culture in teams, which must be considered for computer-aided sprint support.

Research Question 3

How is the utility of computer-aided sprint feedback on socio-technical aspects concerning the understanding for team dynamics in agile projects?

The third research question of this thesis concerns the practical utility and perceived added value of computer-aided sprint feedback in agile software projects. The central focus of this question is to disclose whether the complementary sprint feedback on the socio-technical aspects can objectively or subjectively support the practical understanding in agile teams for arisen team dynamics in the projects.

1.3   Design Science Research

In this thesis, the concept of computer-aided sprint feedback on socio-technical aspects is similar to information system research (ISR) activities. The concept was designed and prototypically realized in iterative refinements using design science research (DSR) [94, 107], which included a focus on the research question defined in Section 1.2.2, particularly for disclosing qualitative and quantitative answers. Figure 1.2 shows the applied design science research in extension of a behavioral science paradigm.

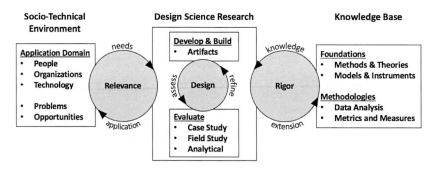

Figure 1.2: Applied Design Science Research, based on [92]

In ISR, the design science paradigm is combinable with the behavioral science paradigm, which is foundational for extending knowledge and understanding application domains concerning relevant environmental opportunities and problems (e.g., including people, organizations, and technologies) [92, 94, 161]. Both paradigms are considered in this research with the purpose of identifying and creating new artifacts (e.g., feedback assets for the communication behavior and mood courses) that support awareness and understanding of team dynamics in agile software projects through supplementary feedback support on socio-technical aspects in sprints. Moreover, DSR was used for prototyping the computer-aided sprint feedback for agile development environments, which integrates methods for capturing socio-technical data and automated feedback processing.

1.3.1   The Relevance Cycle

Modern software development depends more than ever on teams and the people behind them [47, 111, 145]. Software projects that involve agile methods benefit from highly volatile processes that enable teams to react faster to customer demands through manageable release sizes in short development iterations. However, this is not always conducive to a team's socio-psychological well-being [25, 239]. The focus is generally on steady development performance and accurate estimation, often measured using process measures such as team velocity (i.e., scheduled versus completed features) [63, 85, 111]. Problems not only relate to technologies or processes but also to socio-technical dependencies in projects, which makes teams vulnerable to sociological issues, such as dysfunctional communication structures, a hostile atmosphere, demotivation, and dissatisfaction [238]. This applies to both conventional software projects and agile projects. The difference is that in agile projects, the volatile nature of the process facilitates the identification of negative habits earlier through a holistic consideration of socio-technical dependencies in sprints before sociological team issues lead to long-term problems due to a lack of awareness.

The relevance cycle connects the contextual environment of socio-technical aspects of agile software development with the activities of DSR. As a practical example, understanding dysfunctional communication structures in agile teams requires a consideration of the different roles and interests of members, which are shaped by the relevant capabilities and characteristics of people and organizational workflows, the agile practices applied, and the information technologies used during software development. The objective is to understand and identify essential socio-technical dependencies and information-based problems in agile software projects. To this end, it is necessary to derive new and innovative feedback assets for characterizing and understanding team dynamics in sprints based on practical needs [62, 122]. Such assets can facilitate understanding of mutual influences, especially human factors, which are relevant for most projects. Observational studies, surveys, and interviews support the identification of environment-related problems and opportunities for building and evaluating computer-aided sprint feedback artifacts concerning team dynamics in agile projects [128, 133, 207].

1.3.2   The Rigor Cycle

The rigor cycle links design science activities with a knowledge base that comprises interdisciplinary theories, frameworks, instruments, constructs, models, methods, and measures (e.g., from software engineering, data science, and social and organizational psychology) relevant for this research [92]. Moreover, the rigor cycle involves the integration of previous knowledge, experience, and foundations from other disciplines into the system-aided sprint feedback concept to ensure innovation and scientific rigor. For example, previous experiences and observational studies are helpful for determining or adapting sociological measurement methods capturing (e.g., team communication and mood data) [139, 207].

Understanding team dynamics in sprints through computer-aided feedback depends on several interdisciplinary foundations and methodologies that systematically capture and analyze the underlying socio-technical aspects of agile software projects. For example, Goal Question Metrics is a well-known process for identifying goal-oriented metrics even across different disciplines. It can make team behavior quantitatively assessable based on sociological measures in conjunction with standard agile process metrics [17, 178, 207].

In social science, network analysis is a common method for characterizing team communication structures. In data science, analytical methods and machine learning are useful for descriptive and predictive data characterizations (e.g., socio-technical dependencies in sprints).

1.3.3   The Design Cycle

At the center of DSR is the design cycle for the development, evaluation, and refinement of artifacts (feedback assets) [92]. The development of the feedback assets is based on the information needs and opportunities identified during the relevance cycle and is aimed at enabling agile teams to understand the team dynamics in agile software projects (e.g., changing communication behavior over time). The knowledge base developed during the rigor cycle provides the foundations and methods necessary to practically realize the feedback asset (e.g., systematic capture of socio-technical data and preprocessing, followed by descriptive and predictive analytics and incorporated information visualizations as an automated routine).

The behavioral science paradigm is foundational for ISR, as it concerns the confluence of people, organizations, and technologies [93, 94, 161]. For example, agile development teams often lack time to follow up on problem investigations or to develop extended feedback mechanisms due to meager resources and workload estimates in sprints [61, 133].

Consequently, the computer-aided feedback assets in this work are aimed at information simplicity, knowledge supplementation, and fast availability during development processes [166, 233, 239]. For the first functional prototype of the computer-aided feedback support for understanding team dynamics in sprints, 32 observational studies of academic software projects and three industry case studies were involved during different maturity stages of the design cycle. The student software projects were primarily concerned with prototypical design implementations. These are useful for assessing more accessible data acquisition methods and gradually implementing visualizations for the feedback assets without cost-expensive retries in case of malfunctions. The objective is the iterative development, functional assessment, and refinement of the individual computer-aided sprint feedback assets (e.g., retrospective, predictive, and exploratory modules) [134, 136, 139].

The DSR outcomes are feedback assets that support an understanding of team behavior patterns (e.g., a retrospective feedback module that descriptively summarizes team communication behaviors or another that reveals mood patterns) in sprints. They have been qualitatively and quantitatively assessed and refined during their use in different cohorts.

1.4 Action Research

The DSR conducted for this work followed a generalized design purpose for agile development environments and did not result in or assess any client-specific feedback artifacts [107]. Consequently, action research was applied to independently evaluate and improve the utilization and utility of the computer-aided feedback concept with the help of two newly initiated agile development projects from the industry (unknown development conditions) [11, 183].

Action research enabled an evaluation of the applicability, utilization, and utility of the computer-aided sprint feedback in the field. It extended the observational studies of the DSR through a reaction-based methodology aimed at assessing and improving client-specific needs and problems directly (e.g., unusual development behaviors that impact the functionality of the computer-aided feedback). The action research conducted for this thesis involved cyclic qualitative assessments to identify usability and utility deficits in the ongoing projects and support quality-focused improvements based on team reflection on practical problems, joint action planning, and subsequent interventions, followed by assessments of the changes made.

1.5 Contributions

This work makes the following contributions, which can be used as a basis for future research on socio-technical aspects in software engineering:

(1) An overview of relevant socio-technical aspects and practical measurement methods that support an understanding of team dynamics in agile projects.

(2) A concept for exploratory analysis of socio-technical sprint dependencies and visualization through force-directed network graphs.

(3) A feedback concept for computer-aided sprint characterizations and visualizations that considers descriptive and predictive team communications, interaction networks, positive and negative mood affects, and performances.

(4) A practical evaluation of the use of the computer-aided feedback concept in industrial software projects with qualitative and quantitative results.

1.6 Structure of the Thesis

The structure of this work is shown in Figure 1.3. Chapter 2 presents the fundamental information necessary for understanding the feedback concepts. Chapter 3 covers related work on teamwork in software projects, model-based knowledge support, and feedback mechanisms in software development. Chapter 4 presents the concept for the systematic capture of socio-technical data and describes both the process of identifying relevant metrics and capturing methods. Chapter 5 describes the central concept of this thesis, which involves automatic processing, descriptive and exploratory analysis, the development of prediction models, and visualization of computer-aided sprint feedback. The chapter closes with qualitative and quantitative observations from the first technical feasibility study. Chapter 6 describes two industrial software projects that were used to evaluate the applicability and practicability of the computer-aided feedback concept in practice. Chapter 7 summarizes the work and answers the three central research questions. It also discussed the study's limitations and avenues for future research.

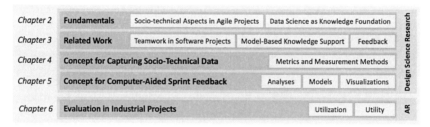

Figure 1.3: Structure of the Thesis

Chapter 2

Fundamentals

The present work focuses on computer-aided feedback to support the team understanding and characterization of team dynamics in agile software development. The data collection and processing in the computer-aided feedback concept described in Chapter 4 and 5 require some fundamental understanding and knowledge for the socio-technical aspects in agile software development, as well the statistical and model-based methods (e.g., descriptive and predictive data analytics) used to characterize team behavior in sprints.

2.1 Socio-Technical Aspects in Agile Software Projects

Traditional software development often involves self-managing professionals with high individual autonomy, but low team autonomy [170, 171]. In contrast, agile software development requires a distinct individual and team autonomy, in which human factors and the behavior of developers determine team success [53, 170, 172]. A lack of support and reduced external autonomy are significant barriers for self-organizing teams, with implications for development teams and managers [170].

Agile teams' social natures depend on individuals' cultural support and social competencies during software development [238]. High character diversity on a team can influence the collaboration culture and familiarity in teams and thus how people behave, organize, and perform [203, 219]. Understanding behavioral patterns can facilitate finding explanations for team dynamics in projects, such as, alternating development performances or changing communication structures) [170, 173].

Curtis et al.'s [53] layered behavioral model abstracts the affect range of individual behaviors in projects and organizations, shown in Figure 2.1. This model demonstrates that performance, commitment, and related social interactions also adduced individuals' emotions, all of which are relevant factors for holistically understanding team dynamics in projects [62, 122, 179].

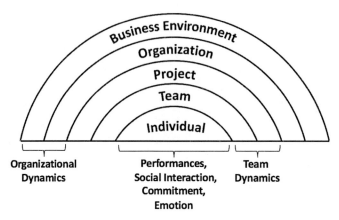

Figure 2.1: Layered Behavioral Model, cf. [53]

In practice, interpreting team behavior based on social and technical aspects is a significant, often challenging task [128, 137]. Insufficient time, technology, and expertise can hamper the consideration of human factors during the development process [133]. Therefore, this research focuses on supporting teams during projects to create a broader understanding of social-technical dependencies and behavioral patterns (e.g., through computer-aided methods and extended feedback mechanisms). The goal is to establish an understanding that does not require significant extra effort or expertise. The following subsections describe the socio-technical aspects associated with team dynamics in agile projects.

2.1.1 Dynamics in Development Team

Team dynamics are behavioral patterns of teams that determine development performance and, therefore, the success of agile software projects [62, 111]. Behavioral changes arise over time for various reasons (e.g., approaching release dates or firefighting situations). A natural sense for such changes involves social dynamics based on people's central involvement in the development process [54].

Understanding team dynamics requires systematic methods for capturing and characterizing typical behavioral patterns in ongoing projects. Additionally, the latter process can be employed to examine temporary dynamics, or the detection of anomalies [122, 169]. The research context of this thesis concerns agile methods and practices that align with people-centered development activities [47, 238]. The foundation for agile development environments and team behavioral aspects (e.g., team performance, communication and interaction, mood, and commitment) is provided in this section.

2.1.2 Agile Development Environment

Agile software development follows the idea of continuous improvements, accomplished through one of the strengths of the agile practice: team communications and customer involvement, incremental releases, self-organizing teams, and intensive team collaboration [145, 184]. Teams and customers often benefit from efficient communication methods (e.g., those that support quick information and progress updates). Beyond this, malfunction or false product development can be rapidly addressed. Moreover, knowledge sharing in teams endorses the effectiveness of members' coordination, and future decision-making [40, 120, 140].

In Scrum, knowledge, and information are regularly communicated in several meetings (e.g., sprint planning, daily Scrum, sprint review and retrospectives) [59, 69, 172, 209]. These meetings constitute an information-rich event where expertise from the past development weeks is applied through discussing ideas, making decisions, and examining what factors will support improvements [59]. For example, in retrospective meetings, a team seeks to identify what went well in the last sprint and what did not to derive improvements. In addition, feedback from team members and customers is a valuable information source and supplements an individual's perspective with the perceptions or experiences of others [47, 132, 239]. Figure 2.2 depicts the sprint cycle and agile practices of Scrum.

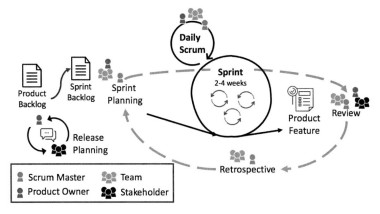

Figure 2.2: Scrum Development Cycle with Agile Practices, cf. [59]

Daily and Weekly Scrum are status meetings held at their respective intervals, each lasting for about 15 minutes. In these meetings, every team member individually reports their accomplishments since the last meeting, including obstacles. Additionally, the targeted development tasks are noted [100]. In Scrum meetings, team members are encouraged to interact and share their work-related knowledge, gaining insights from the rest of the team, especially useful when team members listen to each other as they share [44]. This meeting type facilitates a regular communication base that allows the team to identify and manage development obstacles jointly. Furthermore, it promotes team-oriented decision-making, focusing on all of the members' tasks [45].

Release and Sprint Plannings are held to share knowledge on software requirements and the product domain between the customers and the developers. The
plannings bring both sides together to discuss what will be done in either the next
release or subordinate sprints [50]. Planning meetings aim to break down initial requirements into a manageable set of development tasks to be accomplished within
the next iteration. Related discussions about the product features include estimating the development efforts to be handled by the team [45]. The developers' estimations define future sprint tasks, corresponding to the number of work hours
available in the upcoming sprint, the development velocity achieved in the previous iteration, and the customer's feature prioritization.

Sprint Reviews are conducted at the end of each sprint. In these reviews, the product owner, stakeholder, and the team together discuss what was developed during the last increment [45]. The sprint review targets product-oriented feedback.
Customer change requests and incomplete development tasks are pushed into the
backlog and reprioritized for the next sprint [186]. The time it takes for a sprint
review can significantly vary based on the complexity of a product, the underlying
tasks, and problems that arise [100]. Therefore, small reviews during an ongoing
sprint are common practice to make stakeholders less stressed.

Sprint Retrospectives take place at the end of a sprint. Retrospectives happen behind closed doors, which means that only the team participates in this meeting.
This session facilitates identifying any success factors or obstacles of the current
development process, including social tensions or conflicts. Thus, they internally
reflect on the latest sprint, discussing what worked well, what was not optimal,
and what could be improved [186]. Thus, retrospectives are a proper meeting type
for discussing issues and obstacles experienced by the team during the development iteration. Early recognition of problems supports finding countermeasures
and making adjustments within the project life cycle [150].

These Scrum meetings provide high social accountability, raise the quality of communication, align the team's focus for development activities, and cultivate team
spirit in projects [63, 238]. However, these meetings also primarily focus on sprint
performance and progress metrics (e.g., velocity, estimation error, and other metrics summarized in Appx. A.1). The overview shows frequently used metrics in the
context of agile software development, while human factors or sociological aspects
related to team behaviors are usually not considered. It limits the holistic understanding in teams for socio-technical dependencies and arisen team dynamics.

2.1.3   Tracking Team Performance

The most widely used agile method in practice is Scrum, in which teams benefit
from reliable, lightweight information to disclose development performance over
time [142, 237]. Tracking performance changes allows timely adjustments when
problems arise. A global survey conducted in 2013 revealed that 50% of all agile teams experience difficulties capturing their performances, limiting their im-

provement potential [63]. Metrics with low meaningfulness have proved to increase dysfunctional team behaviors [85]. In this context, advanced and or simple team performance measures usually originate from story points, burndown charts, and accomplished velocity in sprints, which are consistently available in standard progress tracking management systems (e.g., $Jira$, $Trello$, $Asana$) [63].

Story Points describe the estimated effort of development tasks. During sprint planning, a team identifies a simple development task (e.g., story, bugs, etc.) and jointly assign an arbitrary number of points based on expected effort. The other functions of the sprint backlog are then estimated relative to the reference, which is commonly done using the technique known as *Planning Poker* [50].

Burndown Charts support teams to keep track of development progress, based on a linear distribution of the total estimated story points and the number of working days within a sprint. Burndown charts are daily updated so that the trendline (offset) indicates the estimated incomplete story points at the end of Sprint (estimation error). A burndown chart assumes a stable development performance every day. In contrast, an actual daily progress line depicts the completed story points versus the linear reference (ideal line) according to the number of past working days [63]. Thus, the daily progress status becomes more transparent, enabling a team to track actual versus estimated performances and complete the sprint goals in time. Moreover, burndown charts are consistently available in standard progress tracking management systems. Figure 2.3 depicts a non-linear example for the development progress based on the initial planning.

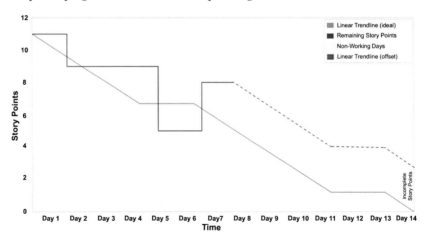

Figure 2.3: Development Progress in a Sprint Visualized as Burndown Chart

Velocity is a team performance metric often considered in agile software development. This metric is the difference between initial story point estimations from sprint plannings and the actual number of completed story points at the end of a sprint, including adopted work [63, 85]. Velocity charts are frequently considered in Scrum meetings, for example, validating improvement interventions made in a previous sprint. However, this metric is only valid when used by the same team at a constant sprint length because its work prediction is based on the teams' prior efforts [178]. Comparing different teams or varying Sprint duration would require a neutral scaling factor (e.g., logged working hours per task or for each day, and equal understanding for the value of one story point). Systematical captures of sociological factors in combination with these team performance measures can promote a holistic understanding of socio-technical dependencies in Sprints, motivating even good teams to become better [59, 133, 188].

It is essential to encourage appropriate metrics that are congruent with the objectives of understanding teamwork and behaviors in agile development teams [85]. For example, several adaptions of the initial velocity metric enable comparable performance characteristics that supply different forms of progress [63]. Moreover, these adaptions do not cause disruptions in teams' daily development tasks or performance because these factors are systematically tracked and processed by the standard management systems mentioned above. Table 2.1 lists a set of helpful velocity adaptations that describe different aspects of teams' development performance in sprints.

Table 2.1: Team Performance Related Velocity Metrics, based on [63, 85]

#	Metric	Definition
1	Work Capacity	The Sum of All (Un)finished Work Reported During the Sprint.
2	Focus Factor	Velocity ÷ Work Capacity
3	Percentage of Adopted Work	$\sum$ (Original Estimates of Adopted Work) ÷ (Original Sprint Forecast)
4	Percentage of Found Work	$\sum$ (Original Estimates of Found Work) ÷ (Original Sprint Forecast)
5	Accuracy of Estimation	1 - ($\sum$ (Estimate Deltas) ÷ Total Forecast)
6	Targeted Value Increase	Current Sprint's Velocity ÷ Original Velocity
7	Win/Loss Record	Sprint was Successful only if: a) At Least 80% of the Original Sprint Forecast has been Accepted. b) Found + Adopted Work was $\leq$ 20% of the Original Sprint Forecast.

Metrics relevant to teams' development performance should be more visible because too many unforeseen tasks reduce a team's sprint velocity, thus increasing technical debts. A qualitative interview study revealed that teams desire a holistic understanding of what is happening [178]. With this in mind, alternating team performances in sprints should facilitate finding explanations. However, not considering team behavioral aspects leaves interpretation gaps regarding the arisen changes [178].

Therefore, it is essential to systematically include socio-technical aspects during the sprint activities (e.g., computer-aided feedback). Interpretations of team dynamics in sprints through systematically captured sociological measures (e.g., emotions, satisfaction), combined with objective performance metrics from progress-tracking management systems, increase awareness and understanding of effects [62, 119]. The computer-aided sprint feedback concept described in Chapter 4 and 5 builds upon this ideology and involves both subjective and objective measures.

2.1.4 Communication Structures in Development Teams

Functional team communication and organizational structures allow a steady information flow relevant to the resulting team performance and effectiveness in agile software development [5, 49, 213]. In contrast, indirect and long communication paths amplify miscommunication, especially in distributed teams, thus endangering project success [34, 44, 52]. Moreover, the definition of software project success shows strong parallels with team success aspects from group theory according to Pinto et al. [188]. Both forms of success depend on the following outcomes:

- **Team psycho-social outcomes** refer to the experienced friendliness, support, positive feelings, acquired knowledge and skills, enjoyment, and sense of pride and added value of participating in the project [188].

- **Software project task outcomes** refer to the estimated tasks and resources and optimal performance in terms of delivered software [89]

Consequently, understanding team communication requires consideration of psycho-social and task-related outcomes in sprints. The reason is clarified through the term *social presence*, which describes the feeling that arises among team members when a sense of open communication with other group members exists [214]. Thus, it matters how teams interact based on different communication channels (e.g., face-to-face interaction in teams causes a distinct perceived social presence than indirect contact by email) [207].

Furthermore, understanding team dynamics in agile software projects is crucial because the volatile development iterations depend on intensive team collaborations and self-organizing structures that support brief communication methods [122]. Teams must react swiftly, for example, in cases of task-related problems or customer reprioritization during the short development iterations [145, 184]. Communication of status information, shared experience, and perceptions can help all members more efficiently recognize and adjust dysfunctional structures (e.g., social conflicts, information availability, etc.) [130, 139]. Therefore, the social perspectives of individuals are essential in practice for agile software development teams [108]. Newly formed agile teams must reach steady communication structures promptly.

However, the more team members, the more difficult it is to manage communications and information flows without losses. Related studies have found that a team's size affects both effectiveness and productivity [4, 145, 156]. For example, small teams ranging from three to seven members have reached efficiency peaks, but team sizes beyond nine begin to decrease the efficiency [4]. Network graphs can support each member's awareness and cognitive understanding of the current state of affairs and identify adequate adjustment opportunities [4]. Figure 2.4 presents communication structures as a graph with a growing number of communication paths, which correspond to team size and the assumptions that everyone communicates to everyone [4].

The graphs underscore that communication networks rapidly grow in complexity, making it challenging to investigate exclusively dysfunctional connections.

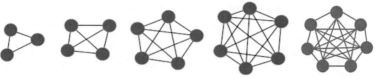

Number of maximal possible Channel with n-Developer = (n x (n - 1) / 2)

Figure 2.4: Complexity Growth of Communication Networks [145]

In agile software development, teams are encouraged to follow a decentralized decision and communication structure. In such a scenario, no single person wields exclusive authority; instead, each member independently makes decisions. Moreover, decisions are made as a team [171]. Nevertheless, dysfunctional organizational structures can arise, such as when they involve maverick activities or introverted or extroverted personalities. These factors often come with higher risks for information loss within a team [212]. Group thinking should guide an agile project and reduce dysfunctional tendencies, including maverick or loner activities because every member contributes to the project's success [4, 39]. However, certain behaviors can endanger the base for the lived social culture in agile processes. Figure 2.5 illustrates an example of centralized and decentralized team communication structures corresponding to the red nodes.

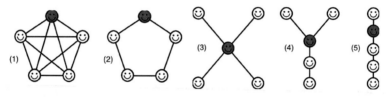

Figure 2.5: Centralized and Decentralized Communication Networks [15]

In the shown example, a complete communication network (1) with five vertices is denoted as K_5. This means that a team communicates through all possible channels ($\frac{n(n-1)}{2}$) so that each developer will cross-communicate with all other team members. Agile teams benefit from short communication ways, enabling rapid information sharing without the constraints of passing central communication nodes or forwarding delays. In contrast, a fully connected communication structure is functional only in small groups because the overhead of possible communication channels increases exponentially with each additional member. It is more challenging to keep everyone informed on larger teams [15].

However, like the "star" structure (3), centralized communication networks can be an efficient communication strategy, considering different roles and restrictions in teams or development workflow. An equally distributed information flow on a team is ensured with a decentralized communication network that handles a shortly omitted member. If a central team member, e.g., the red node in (3), falls apart, the entire communication network stops working.

The characterization of teams' decentralized or centralized communication structures is not the only relevant factor for understanding team dynamics in sprints. Due to software developers' social natures, maverick structures and substructures sometimes manifest over time [212]. Structures involving subgroups or loners, like the two shown Figure 2.6, increase the risk of communication barriers for the information flow in teams. Additionally, such structures limit opportunities for sharing experiences because of member's expertise and knowledge isolation.

Therefore, it is essential for the culture and success of agile teams that such structures to recognize such structures at early stages to prevent more significant harm [62]. Figure 2.6 shows an example of such dysfunctional communication structures. The computer-aided feedback concept described in Chapter 5 considers the communication structure and intensity in teams, along with commonly used media channels during Sprints. Social network analyses (SNA) are used to quantitatively and visually characterize the communication networks as graphs [122, 123, 235].

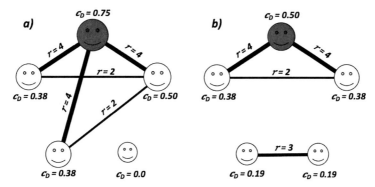

Figure 2.6: Example of Dysfunctional Team Communication Networks

Social networks, by definition, involve a finite set of actors A, and a relationship r, between these actors [123, 235]. In this work's research context, social networks present the set of actors consists of the team members, and the relationship r is given by the perceived communication intensity. The set of all actors in a network is formally defined as follows:

$$\mathcal{A} := \{a_i : 1 \leq i \leq n\} \tag{2.1}$$

The number of actors is defined by the team size n, where a_i stands for an individual team member. The communication relationship between two team members is described by their perceived communication intensity on a five-point scale and formally defined as follows:

$$r : \mathcal{A} \times \mathcal{A} \rightarrow \{0; 1; 2; 3; 4\} \tag{2.2}$$

The relationship between two members is given if the function r returns a value > 0 for both team members. The set of all relationships within a social network is formally defined as follows:

$$\mathcal{R} := \{(a_i, a_j) : 1 \leq i, j \leq n \text{ with } r(a_i, a_j) > 0\} \tag{2.3}$$

The social network analyses in this work are included to identify central members and loners during team communication in ssprints. Therefore, a centrality measure is needed to determine each team member's relative centrality within the team's communication structure based on the members' communication intensities. The degree centrality of a node is defined by its connected edges and respective weights (i.e., the total number of communication paths and intensity by individual members). Consequently, the resulting centrality of an actor a_i in A depends on the total number of connected actors. The centrality measure is formally defined as follows:

$$C_D(a_i) := \frac{1}{(n-1)} \cdot \sum_{j=1}^{n} \{r(a_i, a_j), \text{ where } (a_i, a_j) \in \mathcal{R}\} \tag{2.4}$$

The subsequent visualization through communication graphs supports the cognitive awareness for dysfunctional structures based on the analyzed team communication networks. Moreover, this exercise highlights high centrality in teams and structural risks if central members are omitted shortly. Additionally, this analysis allows further understanding of typical communication behaviors over time, identifying anomalies and inherent team dynamics. Thus, agile teams find support in sprints, more efficiently determining communication problems that hamper or block their workflow and facilitating structural changes or information flows by involving all members, as in the daily Scrum [170].

Besides scheduled team meetings, these communication channels' effectiveness is revealed in the richness of the communication media that teams select [48, 207]. In this context, many organizations also schedule distance meetings using electronic media, such as Skype, Microsoft Teams, and similar software. However, these communication channels' relative values are depend on the situation. That is, a video conversation can be a time-saving communication alternative for a developer instead of face-to-face meetings (e.g., during the COVID-19 pandemic). The communication network fundamentals are needed for the feedback visualization concept described in Chapter 5.

2.1.5 Emotions and Moods in Development Teams

Job performance and satisfaction are intertwined with omnipresent emotions, and moods [196]. In this context, several studies disclosed that happier software development teams perform better (e.g., based on effective problem-solving, social interactions between developers, better quality achievements) [10, 63, 82]. The common problem is that software engineers and managers often use the terms "emotion" and "mood" interchangeably without being aware of these terms' underlying differences, as shown by the bipolar mood model in Figure 2.7. Additionally, they may lack knowledge regarding the relevance of emotional awareness in development performance [77]. Teams may inadvertently run into negative behavioral patterns or miss opportunities to encourage positive actions without suitable mediation. Ben-Ze'ev [28] examined the creation of negative emotions found that people's thoughts about adverse events last five times longer than those regarding positive situations. Therefore, emotions are a critical factor in a team's capacity for rational and ethical thought, including how the members react to (un)pleasant or (dis)engaging situations that occur during the development process (e.g., dealing with an angry customer or positive feedback from a manager) [196].

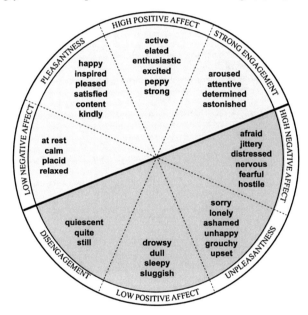

Figure 2.7: Bipolar Mood Model of Positive and Negative Affects, cf. [224, 225]

Emotions are reactions that can turn into moods when the feelings persist even after inciting incidents end. In psychology research, emotions are often classified into a bipolar mood model, as shown in Figure 2.7 [224, 225]. This model separates emotions into positive and negative affects, whereas pleasantness–unpleasantness, and engagement–disengagement on each side represent opposite endpoints.

Robbins et al. [196] described positive affect as a "mood dimension that consists of specific positive emotions such as excitement, self-assurance, and cheerfulness at the high end and boredom, sluggishness, and tiredness at the low end" [196]. In addition, the authors described negative affect as a "mood dimension that consists of emotions such as nervousness, stress, and anxiety at the high end and relaxation, tranquility, and poise at the low end" [196]. Many more emotions exist that do not necessarily fit into exact positive or negative affect classifications. Besides, some of these emotions can hardly be objectively captured in practice, whereas emotional awareness is crucial for development outcomes [77]. This work aims to support the understanding of emotions and moods in the context of team dynamics in agile software development.

The concepts in this work are built upon the well-validated Positive and Negative Affect Schedule (PANAS), an established measure for moods [236]. Although the PANAS is relatively short with its ten positive and ten negative affect items, it is often considered too long in practice. Measuring positive and negative affects must be kept simple, not to cause respondents to experience adverse emotions through tedious self-questionnaires. Schneider et al. [207], and Thompson [225] applied short forms of the PANAS suitable for non-native English speakers to minimize item vagueness and ambiguity. The complexity of PANAS has been reduced (e.g., six positive and six negative affect items) without significantly changing the psychometric properties' reliability. Table 2.2 characterizes the applied survey structure in this research by using PANAS-SF items on a five-point self-assessment scale. Furthermore, the details of this research's systematic data-monitoring concept involving the PANAS-SF measures are described in Chapter 4.

Table 2.2: Adapted PANAS-SF for Mood Affects Survey, cf. [207, 225]

5	always					
4	often					
3	sometimes					
2	rarely					
1	never					
To what extent did you feel _____ during the last week?						
active	1	2	3	4	5
interested	1	2	3	4	5
happy	1	2	3	4	5
strong	1	2	3	4	5
encouraged	1	2	3	4	5
awake	1	2	3	4	5
tempted	1	2	3	4	5
confused	1	2	3	4	5
worried	1	2	3	4	5
angsty	1	2	3	4	5
nervous	1	2	3	4	5
angry	1	2	3	4	5

2.1.6 Performance Satisfaction in Projects

Agile development teams benefit from short release cycles, and increments, which facilitate efficient feedback and close communication with customers [24, 95].

External feedback (e.g., from managers, team leaders, or customers) encourages improvements in team effectiveness (e.g., social skills, nurtured by instilling social and interpersonal skills) [104, 117, 150, 239]. Moreover, the short development iteration enables stakeholders to track the latest development progress every few weeks, ensuring that the product development follows the intended direction. At the same time, teams can build relationships of trust and learn to predict customers' responses over time. In addition, customer satisfaction is often applied as a success indicator in software projects [159, 180, 207]. Hausknecht et al. [88] disclosed that satisfied managers who frequently contact customers positively influence customer satisfaction and loyalty.

However, as previously described, satisfaction is intertwined with emotions and moods, which arise from individual perceptions and are not always correctly interpreted or sensed by teams [196]. Consequently, it is not unusual that customer satisfaction differs from team or manager satisfaction regarding performance outcomes and team commitment in sprints [117, 125, 188]. Supplementary reflections involving individuals' perceived team performance can corroborate solely objective success measures (e.g., requirement compliance, velocity, and defect rate). Such analyses can also supply advantageous insights from different perspectives that promote a holistic understanding of team dynamics that develop in sprints [62, 63, 85].

The adapted version of Kirkman and Rosen's [117] scale of team productivity and empowerment formed the foundation of this thesis. This scale facilitates a systematic assessment of customers' subjective perceptions of team and leader performance [117, 151, 207]. Table 2.3 shows the eight-item team performance assessment. This scale comprises a total of eight team performance properties that can be subjectively rated on a five-point agreement scale (i.e., ranging from 1, "strongly disagree," to 5, "strongly agree") [151, 207].

Table 2.3: Subjective Team Performance Assessment, based on [117, 207]

Perceived Team Performance:

	1	2	3	4	5
5 strongly agree					
4 somewhat agree					
3 neither agree nor disagree					
2 somewhat disagree					
1 strongly disagree					
The communication with the team has always been productive and appropriate.	1	2	3	4	5
The team always seemed motivated and committed.	1	2	3	4	5
The team always seemed organized and prepared at customer meetings.	1	2	3	4	5
The team showed weekly increases in performance.	1	2	3	4	5
The team will achieve its quantitative and qualitative goals.	1	2	3	4	5
The team will exceed the current project requirements.	1	2	3	4	5
The final software product will find few complaints.	1	2	3	4	5
The team develops innovative solutions.	1	2	3	4	5

Teams can reflect their perspectives by frequently obtaining feedback regarding the perceived team performance from different perspectives, particularly from customers and team leaders. For example, strong customer agreement with the statement "The team will achieve the quantitative and qualitative goals" can empower the team when they are unsure if they will satisfy their customer. If the objectively measured development progress matches the customer perceptions, the team will be more confident that their product development will succeed. Moreover, additional feedback and intra-team reflections support group learning and teamwork by focusing on the social factors in the development processes [146]. Additionally, the concept for systematic data-monitoring with relevance to the computer-aided sprint feedback about team dynamics is described in Chapter 4.

2.2   Data Science Support for Knowledge Foundations

In software projects, it is typical for enormous amounts of data to be collected within brief timeframes, often faster than analysts can manually process and interpret. In 1982, Naisbitt said "We are drowning in information, but we are starving for knowledge" and this remains true today [175]. Unfortunately, company, project, and process data are often stored in warehouses and repositories without systematically disclosing hidden information nuggets. However, extracting meaningful information and discovering actionable insights support the knowledge foundation for reliable business and process decisions [71, 148, 191]. Data science is an interdisciplinary field involving principles, processes, and techniques that facilitate understanding phenomena via (automated) data analysis. Figure 2.8 provides an overview of the multidisciplinary nature of incorporating techniques in Data Science that are covered in this work.

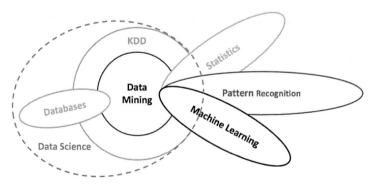

Figure 2.8: Data Science Techniques Applied in this Research, cf. [84]

Additionally, this field extracts valuable information that promotes knowledge, with the goal of improvement [191]. In this context, data analysis methods are increasingly relevant in operational development processes for those wishing to stay economically and technologically competitive in the business market. In agile software projects, turning raw data into insightful information is a time-sensitive activity. In contrast, the short development iterations require fast analyses before the information is outdated at a new sprint. However, manually performed analyses of complex dependencies (e.g., socio-technical aspects) for daily updating input requires advanced data analysis methods that the human mind's processing capabilities cannot match through simple data statistics [116]. Still, it promotes a shared understanding of behavioral patterns (e.g., a relationship between team interaction and development performance) [62].

In software projects, data mining and incorporating techniques such as statistics, pattern recognition, and machine learning (ML) can solve several analysis problems involving the descriptive or predictive characterization of data for knowledge discovery [221, 232]. As a practical example of this work's research context, pattern recognition is applied to detect teams' behavioral (ir)regularities in sprints [169]. In contrast, an ensemble of multiple ML algorithms is involved in predicting team dynamic tendencies for the next sprint [135, 136, 221]. The following subsections briefly cover the data science foundations needed to understand the knowledge base of the computer-aided feedback described in Chapter 5.

2.2.1   Pattern Recognition in Team Behavior

Pattern recognition concerns the (automatic) discovery of (ir)regularities in data, which, depending on data complexity, can be analyzed using simple descriptive statistics (e.g., two-dimension dependencies over time) and ML algorithms (e.g., multivariate relationships in data structures) [31]. In this research, team dynamics refer to behavioral patterns in agile development teams [62, 111]. Understanding team dynamics requires systematic analyses of regular behavioral patterns in ongoing projects to determine anomalous events.

Additionally, recognized problems and anomalous events are usually addressed in retrospective team meetings, making it challenging to explain past inherent behavioral changes resulting from the passage of time (e.g., an increased bug rate midway through a sprint) [180]. Consequently, integrated pattern recognition routines during sprints provide proactive feedback for teams on weak points that could be identified as unusual events that would impact performance. Moreover, this can increase knowledge about dependencies and reduce future risks leading to similar circumstances. This work builds on two complementary pattern recognition techniques involved in the computer-aided sprint feedback concept.

- **Cognitive pattern recognition** involves the detection of patterns in visualized data (e.g., line charts that depict team behavior changes over time to find [ir]regularities) [230, 234].

- **Statistical pattern recognition** involves both simple and complex data analysis methods related to data mining (e.g., descriptive statistics and ML) [79, 116].

Regardless of the recognition technique applied, detecting patterns in team behavior always requires a sufficient amount of socio-technical project data that is consistently collected during the development process. Pattern recognition routines are executed in fixed intervals before data induce possible knowledge, and behavioral rules [187]. The socio-technical data considered in this research are based on a series of data points, ordered in time , such as days, weeks, and sprints).

Line charts present a simple and highly interpretable visualization strategy for time series data [106]. A line chart provides visual clarity, primarily when seasonal trends are targeted over time. A calendar heat map as visualized in Figure 2.9 is a suitable visualization type when searching for patterns (e.g., concerning daily code commits). In such a map, simple behavioral patterns over time may already be cognitively recognizable using fitting information visualization techniques.

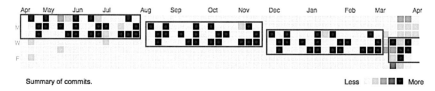

Figure 2.9: Cognitive Pattern Recognition on Code Commits using Heat Maps

The dark green cube presents a high number of commits, whereas a particular working pattern is displayed with seasonal recursion every four months. The time series pattern reveals that the highest commit rates are also linked to a shift of high-performance weekdays every 16 weeks. The earliest time to recognize the high productivity pattern and the weekday shift pattern of the preferred working days is found after only 32 weeks of consistent data. Figure 2.10 gives a simple forecast based on a shift pattern in the previous code commits example. The monthly commit sequence displays a steady pattern, while the forecast of high-performance working days shifts by one week every 16 weeks.

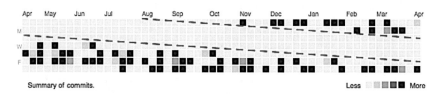

Figure 2.10: Forecast Based on a Shift Pattern of Highly Productive Weekdays

The example includes only a single feature that facilitates detecting simple shift patterns and forecasts related to the team's daily code commit behavior. A more complex pattern recognition (e.g., concerning the relationship between code commits, team mood, and communication structures) requires statistical methods to supplement the human mind's cognitive capabilities [79]. Therefore, cognitive pattern recognition emphasizes involving only a small number of features (e.g., finding independent anomalies or simultaneous peaks between a few features).

Several data science techniques support statistical pattern recognition (e.g., data mining). This allows the creation of (automatic) routines for identifying and characterizing relationships in socio-technical data, thus finding team behavioral patterns in sprints [221]. In this context, Knowledge Discovery in Databases (KDD) is often applied to explore the inherent regularity of data dependencies [71]. The computer-aided feedback concept described in Chapter 5 builds on automated processes for characterizing socio-technical dependencies and statistical pattern recognition for proactive sprint feedback concerning team dynamics anomalies. Additionally, associated information visualizations allow manual pattern reviews for additional cognitive understanding within teams [169].

2.2.2   Knowledge Discovery Support Chain

"Knowledge Discovery in Databases (KDD) is a nontrivial process of identifying valid, novel, potentially useful, and ultimately understandable patterns" [71]. The purpose of involving a knowledge discovery process in this thesis is to accomplish an automated routine for systematically exploring socio-technical data and computation of sprint feedback assets based on descriptive statistics and predictive and exploratory models of previous and future team project dynamics. The KDD process presented in literature is often used for finding and interpreting data patterns, beginning with data selection [71]. This procedure involves the repeated application of preprocessing and transformation steps related to data mining during the building phase, allowing adjustments considering a specified application domain [221]. Figure 2.11 shows the simplified knowledge discovery process used in the computer-aided sprint feedback concept.

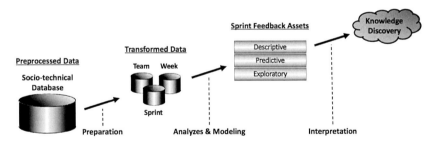

Figure 2.11: Simplified Knowledge Discovery Support Chain, cf. [71]

Preprocessed Data: In this work, the simplified knowledge discovery process considers socio-technical data features as inputs. The preprocessing of collected socio-technical data handles raw data and ensures the cleansing of errors, missing values, inconsistent records, and removing noise or outliers [35].

Transformed Data: The data analyses and modeling methods require transforming the preprocessed data into an applicable input format. For example, the machine learning algorithms of the predictions asset require a standardized *Attribute-Relation File Format* [38]). Moreover, the socio-technical data preparation involves transformation methods to reduce variables or find invariant representations. For instance, daily data of single team members are aggregated or clustered by weeks, and sprints also represent team values. Additionally, several sprint feedback assets require representative team values concerning the information privacy of individuals (e.g., the mood of members).

Sprint Feedback Assets: The descriptive asset covers statistics for the retrospective characterization of central tendencies and variability of the team behavior (e.g., to identify and report team anomalies) [169]. The predictive assets cover an ensemble of machine learning algorithms to disclose team behavior in the next sprint based on previous performances [136]. The exploratory asset covers a system dynamics model with adjustable variables is included for a simulation-based sprint performance exploration [128]. Moreover, it provides exploratory data analyses for characterizing linear and non-linear sprint dependencies, based on the available socio-technical data [134].

Knowledge Discovery: The three assets provide numerically and visualized sprint feedback to support team members' understanding of arisen team dynamics (e.g., as reports). This knowledge is often directly consolidated from sprint reports (e.g., shared usage by teams in sprint Retrospectives) [59, 209]. However, teams and project leaders in the application domain of agile software development must decide whether they want to interpret the computer-aided sprint feedback and discover beneficial actions allowing improvements.

2.2.3 Descriptive Statistics in Sprint Feedback Assets

The computer-aided sprint feedback concept described in Chapter 5 includes on statistical methods for characterization of socio-technical dependencies in sprints and the descriptive determination team behavior course and anomalies (e.g., a significant deviation of the moods). The socio-technical data considered in this work originates from series of data points in time order and mainly relies on ordinal and interval scales (see Chapter 4). The arithmetic mean is generally used for interval scales formally defined as having n represent the total number of all included values from the variable $x = x_i$ [229]:

$$\bar{x} = \frac{1}{n} \sum_{i=1}^{n} x_i \tag{2.5}$$

In contrast, the median can be used for ordinal and interval-scaled variables. The median is frequently used for statistical investigation in social science since most sociological survey data are grouped (e.g., satisfaction is rated based on intervals or ratio scales) without explicit questions for continuous values. The median cannot be used for nominally scaled variables whose characteristics have not natural rankings (e.g., project id's) [221].

The median is formally defined as having n represent the total number of all included values (ranked order) from the variable x [229]:

$$median(x) = \begin{cases} x_{\frac{n+1}{2}} & n \text{ odd} \\ \frac{1}{2}(x_{\frac{n}{2}} + x_{\frac{n}{2}+1}) & n \text{ even} \end{cases} \qquad (2.6)$$

The arithmetic mean $\bar{x}$ and the *median* are primarily used as measures to describe the middle of the socio-technical data variables' distribution according to a single value (e.g., the average of completed development tasks or the typical team mood by days, weeks, and sprints). In this context, neither measure alone is sufficient for describing the value distribution of the variable x over time. Therefore, the variance σ^2 is used to describe the (standard) deviation σ around the mean center. The variance and the standard deviation are dispersion measures that indicate how broadly the investigated data spread from their average [229].

The variance of a population σ^2 is a standard measure in descriptive statistics used to describe the data's dispersion around the mean (i.e., how close every value of the data variable x is to the mean). The variance is derived from the average squared distance from the mean. The deviation mean's square ensures that the variance results are not zero when the deviation from the mean is always zero. The variance is formally defined as follows [229]:

$$\sigma^2 = \frac{1}{n} \sum_{i=1}^{n} (x_i - \bar{x})^2 \qquad (2.7)$$

However, the variance is often used for calculating standard deviation. The measure determines the distance from the mean that is acceptable in a given distribution. In this work, the standard deviation is used in the anomaly detection to identify significant team behavior changes over time (see Section 5.2.1). The square root of variance can determine the standard deviation of all data points relative to the mean. For instance, a small standard deviation in the dataset implies that the data points are close to the mean. The standard variance of a population is formally defined as follows [229]:

$$\sigma = \sqrt{\frac{1}{n} \sum_{i=1}^{n} (x_i - \bar{x})^2} \qquad (2.8)$$

The simple tendency and variability measures involved in this work enable teams to gain feedback and insights about the standard team behavior course. Similarly, computer-aided routines facilitate anomalies' systematic detection based on significant positive or negative course impacts that are considerably beyond the past team behavior tendencies [186]. In this context, an anomaly is defined as an "observation that does not fit the distribution of the data for normal instances" [221].

2.2.4 Machine Learning in Sprint Feedback Assets

Descriptive statistics are considered for the computer-aided feedback concept of this work to determine team behavior courses and variability in past sprints. Model-based predictions of the team behavior in the next sprint allows supplementary advantages concerning the decision-making and planning in ongoing projects [112, 136]. Therefore, foundations from statistical learning are relevant for this work (e.g., supervised learning models). Machine learning (ML) supports predictions based on functional analyses [9]. Furthermore, practical ML goals promote understanding of predictions (e.g., association rules, regression functions). Supervised ML results in mathematical models allowing inferences from historical data [9, 136]. In software development, ML is currently applied to decision-making problems, and performances improvements related to effort estimations, predictions of software failures, and the priority of a reported bug in sprints [20, 237].

Machine learning models built upon different learning categories (e.g., supervised learning, unsupervised learning, and reinforcement learning) [9]. The ML models relevant in this work originate from supervised learning. For reason of completeness, the other two learning types are briefly introduced at the end of this section.

2.2.4.1 Supervised Learning Models

The most widely used ML models are based on supervised learning [87]. Supervised learning addresses either regression or a classification problem, depending on the ML prediction goals. Therefore, a data point relationship is described by a regression equation, such as function $f(\cdot)$, to compute an output y for each input x where the input maps the output best corresponding to the training data. Figure 2.12 shows an example of a learning problem involving training data described by linear and nonlinear regression functions.

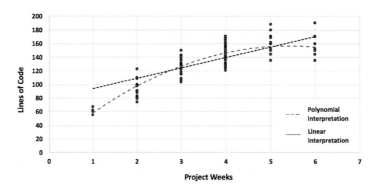

Figure 2.12: Example of Linear and Polynomial Regression Function, cf. [137]

Linear regression is an approach to modeling the linear relationship between the data points of two variables (features) [9]. Regression methods use descriptive statistics to derive prediction functions for linear input and output dependencies between two data features and is formally defined as follow:

$$y = \beta_0 + \beta_1 X + \varepsilon \qquad (2.9)$$

The dependent variable (y) is predicted for any value of the independent variable (x), where β_0 is the intercept (i.e., the predicted value of y when the x is 0) and β_1 is the regression coefficient (i.e., the estimated change of y when the independent variable (x) increases. The estimated error ε determines the variation range of the regression coefficient. A linear regression models resolves the regression line between the dependent and independent variables, if the relationship is linear.

Figure 2.12 shows an example, in which the productivity (LOC) interpretation over time disclosed a positive linear relationship with a strong correlation coefficient $r = 0.74$ and a significance $p < 0.01$ [137]. The coefficient of determination R^2 describes how well the derived regression function fits the training data points between 0.0 and 1.0, whereas a value of 1.0 indicates a perfect fit [9, 87]. In this example, the linear regression only resulted in $R^2 = .55$, so that only 55% of all data points could be mapped adequately using the regression model. A common misinterpretation of a low coefficient of determination is that there is no or only a weak relationship between the features, disproved by the additionally applied polynomial regression using the same data [137].

Polynomial regression is an approach to modeling a non-linear relationship between an independent and a dependent variable with a power greater than one. Polynomial regression is a curve-fitting approach that functionally derives a curve to fit best a series of training data points [177]. The regression example in Figure 2.12 reveals that the derived linear regression model is too restrictive, whereas a quadratic model (polynomial regression model of power two) better solves this problem. In practice, linear regression is applied because of its simplicity, but lack of further investigation (e.g., non-linear analyses) in the case of a low coefficient of determination [194]. Moreover, functional relationships in project data, primarily those involving human factors, can vary over time due to team behavioral changes. Thus, it is essential not to limit prediction models to linearity measures [137, 237].

Various ML algorithms exist, each with a unique methodology for deriving linear or non-linear predictions (e.g., decision trees, logistic regression, support vector machines, neural networks, kernel machines) [13, 87]. However, the training, testing, and derived predictions of supervised learning models are solely data-driven and thus strongly rely on the applied historical projects' data [237]. Figure 2.13 shows the standardized training, testing, and predicting routine for the prediction concept in this work. A shared learning problem involves inferring the function mapping for the input and output based on the training data. An ensemble of ML models is involved for predictive assets of the computer-aided feedback concept described in Chapter 5.

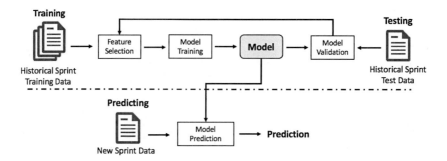

Figure 2.13: Machine Learning: Model Training, Testing and Prediction, cf. [13]

Model Training: The feature selection is a crucial step in the model training activity concerning the model's prediction accuracy. An accurate regression function depends on a significant relationship between the dependent variable y (predicted variable) and one or more relevant regressors (predicting variables) [9, 87]. For example, a regression function that predicts the number of completed tasks in a sprint could be derived considering the number of actively involved developers and their respective working hours as regressors.

Additionally, irrelevant predictor variables without significant relationships to the dependent variable (e.g., the age or gender of the members) influence the prediction accuracy. A manual feature selection usually requires several iterations to identify the feature set qualified as regressors. However, an auto-feature selection process can automate the step (e.g., wrapper-based feature selection) routine through an iterative exploration of different feature combinations, resulting in the highest prediction accuracy [38, 226]. Both manual and automated feature selections aim to produce regression functions with high prediction conformity. A model validation after each iteration allows one to determine whether the changed features led to prediction improvements [9, 13].

Model Validation: Before integrating a trained ML model into practical applications, one should perform a model validation if one was not previously executed during feature selection. Therefore, cross-validation is a widely applied technique for determining the overall classification and regression-based prediction accuracy of an ML model [9]. Additionally, cross-validation is helpful for an ML model selection routine, usually applied to compare the classification or regression performances of different models based on the same training and testing data [221, 226].

Performing a cross-validation in the context of an ML model performance validation requires subdividing the (feature-selected) historical data into two parts: *training data* and *testing data*. While the training data is used to build the model (i.e., to statistically derive the classification or regression function), the testing data is used as input reference data to validate an ML algorithm's prediction performance. The segmentation of data into training and testing sets is often accomplished using *Leave-One-Out Cross-Validation* and *K-Fold Cross-Validation* [9].

- **K-Fold Cross-Validation** can be applied instead of LOOCV to identify whether an ML model is overfitting or underfitting due to the training data [131]. Therefore, the value of K must be greater than or equal to 2. The K-Fold Cross-Validation segments the historical dataset into K mutually exclusive subsets. Each subset is used as a testing set in every iteration, while the remaining $K-1$ subsets are used for training the ML model. This process is iterated through all of the K-folds and results in K-models. The average of all prediction errors during each iteration determines the classification or regression-based prediction accuracy [9]. An example of this K-Fold Cross-Validation with $K=10$ is shown in Figure 2.14.

- **Leave-One-Out Cross-Validation** (LOOCV) is similar to K-fold Cross-Validation with $K=1$. Every data point of the selected features (independent variables) within a historical dataset of size N is reserved as input data (validation) for the model trained by the remaining $N-1$ samples. With N iterations performed, each data point of the selected features has been used once as input data. It allows the comparison between the excluded dependent variables' value and that of the one predicted by the ML model [43, 243]. The overall prediction accuracy results from the averaging error between the predicted outcome and the left-out reference data. Therefore, the cross-validation method considers both the individual training error and the generalization error [131].

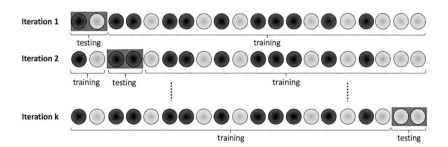

Figure 2.14: Example of K-Fold Cross-Validation with $K = 10$

Model Prediction: Once a supervised ML model is trained, and its prediction performance results in a satisfying overall accuracy, it is deployable to the targeted application context. In the concept evaluation described in Chapter 6, predictions are derived for the following two weeks considering the recent weeks' newly collected sprint data and the historical project data stored in the socio-technical database. After a prediction is derived, the newly collected data are considered in a subsequent learning routine that updates the ML model's training foundation.

2.2.4.2   Brief Introduction of Unsupervised and Reinforced Learning

Frequently applied alternatives to supervised learning models include unsupervised and reinforced learning models [9]. However, both learning categories are not considered for the computer-aided feedback concept for reasons of simplicity.

Unsupervised Learning contrasts with supervised learning in a way that is categorized as self-organized learning, allowing one to detect previously unknown patterns in a dataset without labels (i.e., only input data are considered). The self-organization characteristic enables the model to evolve using probability densities of given inputs (e.g., determined regularities in the data based on clusters) [9, 97]. Therefore, unforeseen prediction outcomes usually require additional analysis to produce valuable insights, which requires greater efforts than supervised learning.

Reinforcement Learning is commonly used in applications whose outputs relate to a sequence of actions (e.g., a recommendation system that requires active confirmations or rejections from the user regarding whether a recommended output matched the expectation derived from a series of given question answers) [9]. Therefore, reinforcement learning models can develop according to an individual's sequence of responses associated with a prediction goal, assessed by the recommended output's goodness based on the reinforcement. Reinforcement Learning requires greater efforts from teams than simple supervised learning predictions concerning a continuous need to confirm or reject actions.

2.2.5   Exploratory Sprint Feedback Assets

The previously presented descriptive statistics and ML models supply the foundations for retrospective and predictive computer-aided sprint feedback concerning team behavior and performances in sprints. However, these information assets focus on how team behaviors and performance change over time rather than why these changes occur. An exploration of team behavioral dynamics and dependencies in Sprints are barely supported by previous techniques. In this context, the foundations of exploratory dependency analyses and system dynamics modeling are relevant to the computer-aided feedback concept.

2.2.5.1   Characterizing Sprint Dependencies

Modeling of socio-technical dependencies in agile development environments promotes a better understanding of team dynamics in sprints [134]. Human factors and underlying phenomena in development teams are often interpreted using empirical studies that disclose valuable insights [16, 173]. To date, however, research has investigated individual practices in isolation rather than as an integrated system [41]. The determination of dependencies often occurs solely by searching for linear correlations, resulting in suboptimal regression models [137, 194].

The exploratory sprint feedback asset in this thesis involves conventional descriptive statistics to detect linear socio-technical dependencies in sprints, together with a lightweight, exploratory algorithm for measuring the *Maximal Information Coefficient* (MIC) in (non)-linear relationships [194]. The MIC's functional property analyzes support, detecting non-linear socio-technical dependencies without relying on the kernel-based operation (e.g., used by Support Vector Machines) [9]. MIC is a measure of the association through a function that rates the strength of statistical dependence between two variables supplementary to conventional Pearson and Spearman correlation analyses for a linear relationship without investigating non-linear characteristics. Figure 2.15 shows a data relationship example that reveals the difference between the MIC's detection (e.g., of sinusoidal relationships), which the Pearson correlation coefficient can not detect.

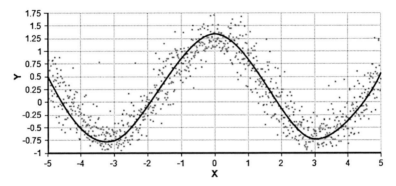

Figure 2.15: Exploratory Relationship Finding: MIC $= .7$ and Pearson $r = .0$

The MIC is based on *Mutual information* (MI), the quantified dependency between the joint distribution of the example variables X and Y and what the joint distribution would be if X and Y were independent. In information theory, MI is a measure of how much information one variable can give about another [141]. Many dependencies can elude linearity measures, motivating the use of other measures of association. The example in Figure 2.15 discloses a linear correlation coefficient $r = .0$ but displayed a high MI, implying a dependency undetected by conventional correlation measures for linearity. MI is non-negative and zero when X and Y are statistically independent [21]. The formal definition of MI is described by each variable's entropy $H(X|Y)$ and the joint entropy of both variables $H(X, Y)$.

$$H(X) = -\sum_i p(z_i) \log(p(z_i)) \tag{2.10}$$

$$H(X, Y) = -\sum_{i,j} p(x_i, y_j) \log(p(x_i, y_j)) \tag{2.11}$$

$$MI = H(X) + H(Y) - H(X, Y) \tag{2.12}$$

The MIC-algorithm discretizes the data onto two-dimensional grids and calculates normalized scores representing the mutual information. The MIC indicates how strongly the two variables are correlated through normalized scores ranging from *uncorrelated = 0.0* to maximal *correlated = 1.0* [194]. For noiseless functional relationships, MIC is comparable to the coefficient of determination (R^2) [194].

The MIC and the underlying MI measures from information theory have been included in this thesis because of their advantages for detecting functional relationships among data features in extensions to solely linearity measures. When summarizing and exploring socio-technical dependencies in sprints, it is essential to holistically investigate data relationships. As previously described in Subsection 2.2.4.1, a common interpretation mistake is assuming that a low linear correlation coefficient or value of zero does automatically imply the non-presence of a non-linear relationship. For the sinusoidal dependency example in Figure 2.15 (e.g., in the case of alternating meeting durations every week), a conventional Pearson correlation would have led to a false-negative interpretation when compared with the extended functional property analyses of MIC. In this context, supplementary visual modeling of linear and non-linear dependency findings (e.g., using networks or hierarchy graphs) promotes cognitive awareness and understanding of socio-technical relationships and the underlying team dynamics in sprints [134].

2.2.5.2   Modeling System Dynamics

The foundations of System Dynamics (SD) originated in the 1950s [74]. An SD model encourages the exploration and discovery of the dynamic behaviors of a system [75]. It presents the decomposition of complex structures into a simple and abstract form. It provides a rational understanding of a project's dependencies and process dynamics (e.g., the socio-technical aspects) [1, 14, 74, 217]. In software projects, several dependencies are involved during the development process (e.g., how an employee's mood affects productivity or how communication structures affect perceived information flows) [14, 134, 158].

Conceptualizing an SD model requires the aforementioned understanding and knowledge of system behaviors to depict the underlying mechanisms in causal loop diagrams, and stock-flow models [8]. In this way, sprint simulations based on known dependencies facilitate the exploration of different scenarios through adjustable parameter settings (e.g., the impact of a higher workload on team mood) that can provide insightful feedback for the next sprint planning. Causal loop diagrams present simplifications of cause-effect relationships and feedback loops in a system [216]. They reflect people's mental models of causation, which help increase understanding based on visualized logical assumptions, reasoning chains, and perception shares. Mainly, they are modeled before a simulation analysis to depict the underlying causal mechanisms that underlie behavior dynamics over time as an endogenous consequence of the feedback structures in a system [192, 216]. Conceptualizing a causal loop diagram is an iterative process in which the number of elements considered is usually modified or extended until the model depicts the perceived structures of the system [2, 14].

An element of a system that indirectly influences itself represents a closed sequence called a feedback loop, or a causal loop [195]. A typical example in SD modeling, known as the *Mythical Man-Month*, is expressed by Brooks's law as "Adding manpower to a late software project makes it later" [37]. An example of the causal loop diagram of Brooks's law is visualized in Figure 2.16.

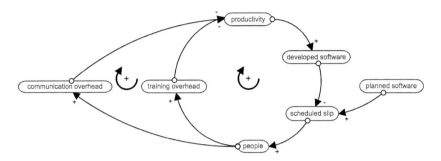

Figure 2.16: Example of a Causal Loop Diagram, based on Brooks's Law [158]

The diagram thematizes the dynamic effects of different time-related hiring conditions during software projects, considering the planned and as-yet undeveloped software corresponding to the remaining duration of the project. With additional training and communication overhead due to late-joining developers, counterproductive effects hamper productivity, thus leading to a delay of the release [2]. The assumed causal relationships between the system elements are expressed through directed arrows, each with a "+" or a polarity sign. The polarity sign characterizes the direction of influence of an element on another, e.g., the negative causal link between "communication overhead" and "productivity" means that with more communications, the "productivity" will decrease. The positive causal link between "productivity" and "developed software" means that increasing "productivity" causes an increase in the "developed software" progress.

The example causal loop diagram of Brooks's law in Figure 2.16 involves two positive feedback loops, each with an even number of negative causal links. Every feedback loop within a system has either a positive (*reinforcing*) or negative (*balancing*) effect and is the algebraic product of the polarity of all causal links in a loop [118]. Reinforcing feedback loops have an even number of negative links, whereas balancing feedback loops always hold an odd number of negative links [1].

Causal loop diagrams are a static representation of the expected dependencies within a system. Stock and flow diagrams are models of integral finite difference equations involving the system's feedback loop structure variables that simulate the system's dynamic behavior. Stock and flow models are commonly used for a simulation-based model that identifies dynamics from the causal loop diagram. Each model is a composition of system variables (i.e., flows, auxiliaries, and stocks), as shown in Figure 2.17.

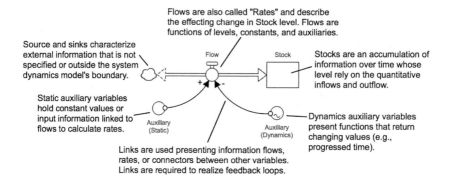

Figure 2.17: Elements of a Stock and Flow Model [158]

A stock represents a quantitative unit of any entity that accumulates or depletes over time [1]. A flow is the rate of change in a stock. There are two types of flows: (1) material flows, and (2) information dependencies. Material flows may only connect stocks [30]. Information dependencies must not point to stocks. Stocks have units of quantity, for which connected flow determines the inflow and outflow rates. With this, the quantitative unit held by stock changes over time, depending on its incoming and outgoing information or material flow. The stock and flow model corresponding to the causal loop diagram example is shown in Figure 2.18.

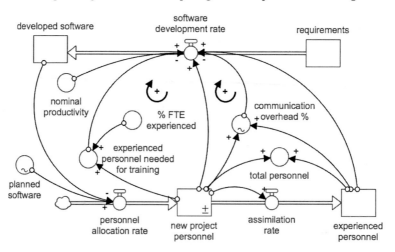

Figure 2.18: Example of a Stock and Flow Model, based Brooks's Law [158]

Each closed-path causal loop requires at least one stock element to characterize the accumulated level of change. When the development productivity is too slow, in-tuitively, involving additional developers in a project could address the problem.

With this, the *Mythical Man-Month* of Brooks's law primarily targets productivity dynamics over time, which has a relationship to communication and training overheads. At the same time, productivity dynamics depend on whether the developers come with experiences or simply join in the project [37, 158]. Inexperienced people are assumed to require additional training to reach the average development speed of the project group. Brooks [37] reported that such late-joining employees could cause project release delays.

A static stock and flow diagram promotes a basic understanding of dependency structures without the need to run simulations. Nevertheless, it presents only a static model and does not allow the exploration of dynamic system behaviors as they arise over time through the simulations. The quantification of parameters and the formalization of the dependencies' underlying the mathematical equations are significant in turning a static stock and flow model into an executable simulation model [3]. To quantify the model, a modeler has to level and rate equations and estimate parameter values for the involved elements. The following example shows the standard notation involving the inflow and outflow rates for accumulating a stock level at any time t. The stocks are defined through equations as follows:

$$\text{Stock}_{Level} = Level_{t=0} + \int_{0}^{t} (inflow - outflow)\, dt \tag{2.13}$$

The related literature proposes the following five steps for quantifying a system dynamics model [2, 30]:

1. Define the start and end times for the simulation run.

2. Formalize all flow and auxiliary factors, including functions for all links.

3. Define constant values or time-dependent functions (i.e., factors without incoming dependencies).

4. Set the initial simulation level of all stocks.

5. Define the values or time-dependent functions for all factors with outgoing dependencies.

All parameters require reasonable estimations based on general knowledge, data sources, or direct observations [158]. A simulation scenario could be for the stock-and-flow model of Brooks's Law in Figure 2.18, an initial level of "experienced developers" $= 20$ and the "requirements" tasks $= 500$. The *developed software* has an initial value equal to 0, assuming no work was completed upfront. The total number of "project personnel" equals the "new project personnel" plus the "experienced personnel" minus the number of "experienced personnel needed for training" the newly joining developers.

However, not all auxiliaries or flow equations are trivial to formalize. For example, when developers join a project late, additional linear overhead is needed to train the new people to complete a software project in time or finish earlier. The simulation plot in Figure 2.19 characterizes the dynamic effects of different hiring conditions on the development progress and changing project completion times. In the simulation, the new developers take approximately 20 days before reaching a satisfactory experience level. With an increasing team size, the related communication overhead is expressed through a non-linear function ($0.06 * x^2$, where x is the number of people involved). Such overheads also result in adverse effects when developers join a project late because it takes them time to get trained, causing higher communication complexity [2]. The system dynamics model of Brooks's law in Figure 2.19 reveals that the dynamic effects of the different hiring conditions result in changing "software development rates," causing a delay or a shortening of the final project completion.

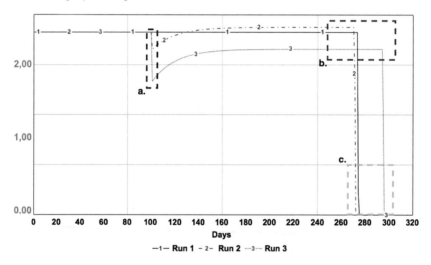

Figure 2.19: Example of Sensitivity Analyses for the Productivity

Sensitivity analyses (e.g., with three runs and different "new project personnel" values at $t = 100$) are performed to investigate the effects (e.g., the "software development rate" related to the training and communication overhead). An addition of five new developers at marker a leads to faster project completion than no joiners, and the addition of ten new developers causes a proportionally high delay in c for finishing all task requirements. In this particular comparison, adding no developers would make it even more efficient to complete the project with a linear estimated duration and human resources. The subsequently affected "software development rates" in b explain these productivity gaps and changed project completion time.

Chapter 3

Related Work

This thesis involves theoretical and practical methods from software engineering, data science, and social and behavioral science. Related work from these fields built the foundation for this works' computer-aided sprint feedback concept concerning agile software development team dynamics.

3.1 Teamwork in Software Projects

Several studies in the fields of software engineering and social and behavioral science describe the role of human factors and the associated team dynamics in software development processes. Related work on team behavior and teamwork in software projects is fundamentally important to understanding how development teams communicate in practice and how existing methods capture these teams' sociological and behavioral aspects.

Cockburn et al. [47] addressed the effects of working in agile software development and the role of people involved. The authors previously investigated the importance of teamwork based on individual and team competencies to capitalize on everyone's unique strengths and overcome vagaries related to the advantages and challenges of using agile methods. Moreover, the authors highlighted the importance of managers following the agile spirit in software projects (i.e., emphasizing human factors, including amicability, talent, skill, and communication). They also reported that agile teams could benefit from members having different personalities in a co-located working environment, which could positively affect team communication and organizational structures [48].

Moe et al. [170] investigated the interrelations between human factors and effective teamwork in the context of agile software development. The authors conducted multiple field studies with professionals in agile development environments to better understand the nature of self-managing teams and the challenges and influences for effective teamwork [172, 173]. Their extensive fieldwork focused on human sensemaking and how the people involved perceived the mechanisms of teamwork during the development process. Moe et al. [170, 173] observed specific barriers to trust and effectiveness in teams depending on orientation, leadership, and project coordination. Moreover, the study found obstacles for highly specialized developer skills and the division of development tasks, and a lack of systems that support team autonomy. The authors found that understanding the operative behavior of teams and different developer personalities, skills, and ambitions within organizational structures allows improvements to areas such as reducing estimation errors [172].

Hoegl et al. [98] investigated the quality of teamwork and how it is measured. The authors developed a teamwork quality model concerning the relationship between team performance and communication, coordination, the balance of member contributions, mutual support, effort, and cohesion (rated by team members, leaders, and team-external managers). They conducted individual interviews and utilized a fully standardized questionnaire in which respondents were asked to evaluate the team's properties and behaviors as a whole. The results related to data from 145 development teams [98]. The authors also noted a strong correlation between team performance and individual members' success (i.e., work satisfaction and learning). Moreover, the perceived team performance varied depending on the rater's project role (i.e., managers compared with leaders and team members).

Herbsleb and Mockus [91] conducted an empirical study to investigate differences in development cycle times between co-located and globally distributed development teams. The authors explored possible reasons why distributed networks are less effective and accomplish tasks less quickly than co-located teams. The authors used quantified source code data from change management systems and a survey to capture team experiences (e.g., workload pressure and team affiliation). Their observations confirmed that distributed deployed work items (e.g., new functionality or problem fixing) took about two and one-half times longer to complete than similar tasks by co-located teams [91]. In addition, the researchers found that people located 30 meters apart communicated no more frequently than globally distributed teams. The study showed that development delays were caused by considerable differences in the communication network's size, leading to a restricted flow of information across sites (e.g., difficulty finding people and reducing the likelihood of obtaining helpful information). The study found instances of reporting sociological problems such as a perceived low "teamness," which led to some work conflicts related to team support in high workload situations. While the use of some communication and media channels seemed to restore this lack of communication, the empirical study found that increased usage of telephone and e-mail communication seemed insufficient to deliver the expected information [91].

Schneider et al. [205, 207] used longitudinal studies with student software projects to investigate the relationships between media, mood, and meetings in teams and how they related to project success. The authors adduced data showing that indirect communication and inappropriate media in distributed teams could create a perceptual distance, intensifying team members' feelings of isolation and distance from each other. Moreover, the authors investigated whether the perceived distance could also impact individuals' moods in teams and, consequently, the projects' success. The study was conducted using interdisciplinarity foundations and perspectives from software engineering, organizational and social psychology. The authors used statistical correlations, and non-parametric tests for quantitative empirical analysis of opinion polls concerning communication networks, mood, workload dependencies, technical process measures, and meeting behavior-based video coding schemes [205, 207]. They quantified communication behavior using the so-called FLOW distance as a measure for indirection in information flow. The authors captured changing moods and arising social conflicts using positive and negative affect scales from psychology [236]. They identified significant team behavioral patterns between communication structures, mood, and social conflicts, which led to the conclusion that early indicators of isolated team members or excessive participation variance concerning the teams' mood, communication, and collaboration can help to avoid later problems, and, thus, should complement solely technical process measures [130, 205, 207].

O'Connor and Basri [179] conducted a study concerning the effect of team dynamics on the software development process in small companies. The authors focused on teamwork issues and the often less considered team dynamics impact during the development process. Following other research, the authors considered the term "team dynamics" as patterns of interactions among team members that determine the development's performance and success [62, 122, 179]. The authors used interviews and survey studies to capture individuals' attitudes and experiences and explore behavioral perceptions, values, and feelings. Their research involved quantitative analyses that led to several conclusions, including that positive team dynamics make teamwork more productive and promote a better working environment with satisfied and fulfilled employees. The study was highly influenced by team structure [179]. Moreover, the authors found that good interpersonal skills, working in close proximity, strong social relationships, and a willingness to share opinions and ideas frequently during the planning and strategic meetings impacted the software development process [179].

Klonek et al. [122] described a quantitative observational method for capturing and visualizing communication dynamics in team meetings. They found that systematic analyses of team interactions can help understand the temporal dynamics of team behavior in certain situations. Capturing team dynamics in a fully-situated context without observational biases can be challenging, and it can require advanced methods beyond standard self-reporting measures. Hence, the authors developed a browser-based communication analysis tool that captured interactions in group and team meetings. This analysis tool was based on coding schemes that structured the meeting observations based on an Interaction Process Analysis (IPA) coding system for video and audio recordings or via live observations [122]. The interaction coding tool supported consecutive visualization and feedback options for displaying tracked team dynamics. It included an interface for conducting advanced statistical analyses on effective team processes. The practical application revealed the tool's supporting benefits for instantly available feedback on team dynamics at the end of meetings, allowing the team to understand how specific team activities had shifted during the session [122].

3.2 Model-Based Knowledge Support in Software Projects

Model-based approaches are frequently considered when organizations seek improvements in development processes (e.g., to maximize team performance) in ongoing software projects. Interdisciplinary techniques from data science support knowledge discovery in project data with computer-aided feedback research (e.g., machine learning and exploratory models). Given this, it is not surprising that several related studies on software engineering have already focused on integrating model-based knowledge support with development processes for practical team support. Only a few studies focus on understanding team dynamics.

Jørgensen [113] described a study comparing several prediction models (e.g., simple regression, multiple regression, neural networks, and pattern recognition) for estimating the different software maintenance efforts in projects. The author found that for managers, it is essential to evaluate and plan the costs and benefits of maintenance tasks accurately through experiences and computer support. The prediction models were trained and validated using the data of 106 maintenance activities. Each model's prediction accuracy was compared with the others and with the managers' experience-based estimation. The most accurate predictions were achieved by applying models based on multiple regression and pattern recognition, which reached an accuracy of only 60%. This was explained by differences in the environments from which the data was collected. Jørgensen [113] concluded that a prediction model should not replace expert estimations based on personal knowledge, which are not included in a formal model. Instead, Jørgensen et al. [114] suggest using prediction models as complementary instruments to support expert estimates and to analyze the impacts of processes and products.

Vetro et al. [232] presented a study in which the authors described the combined use of software data analytics and teams' feedback to improve sprint estimations in Agile software development. The authors used user story information from the project management system JIRA to analyze the variance in sprint estimations and team feedback. They used this information to understand the causes of inaccurate estimates in sprints, such as underestimated tasks and nonconstant complexity ratings. The feedback elicited in the team meetings and complementary visualizations of regression-based user stories analyses led to significant changes in the estimation process with an improved estimation accuracy between 10% to 45% [232]. The changes included, for example, an analogy-based estimation and retrospectives on the accuracy of past Sprint estimations.

Batarseh et al. [20] introduced a method that supports predicting software failures in the next agile sprints using analytical and statistical methods such as regression-based modeling. Their method was data-driven by continuously measuring Mean Time Between Failures for software components during sprints and predicting errors with statistical confidence. Based on previous records, the error forecasts were based on a regression model that estimated where and what types of software component failures are likely to occur.

Wolf et al. [242] described a study about the prediction of build failures in software projects using social network analysis on team communications. The authors analyzed five project teams' communication structures using social network measures (e.g., density and centrality). They then systematically matched the teams' coordination outcomes with their code integration processes of successful or failed builds related to artifacts that included build results, work items, contributors, and comments. Finally, they applied communication structure measures into a predictive model based on a Bayesian classifier that indicated whether integration would fail with a recall between 55% and 75% and precision values between 50% and 76%. Their findings revealed that developer communication played a vital role in the quality of software integration and served as complementary planning support.

Forrester [74] pioneered the system dynamics model, a methodology for the abstraction and model-based simulation of dynamics in social systems. System Dynamics was developed for identifying, investigating, and understanding behavior forces in complex systems [76]. The author used quantitative flow models to simulate system behavior changes over time, including quantifiable and interconnected components, stocks, rates, and auxiliary variables modeling the system. Therefore, qualitative modeling builds upon shared mental models and expert knowledge (e.g., causal loop diagrams), differentiating it from other data-driven or statistical modeling methodologies.

Abdel-Hamid and Madnick [2] introduced the principles of dynamic events in software project management. They used multiple industrial case studies with large software projects providing relevant metrics. The authors also used data-based knowledge and subjective experiences to build system models with different stages of complexity. They presented a detailed overview of dynamic modeling examples such as the dependence of productivity on development teams' motivation and even an entire software development process chain. Madachy et al. [158] described the early stages of modeling communication and team issues in this same application context. The authors applied qualitative model simulations to different process dynamics through regular boundary expressions under various parameter settings. Houghton et al. [102] focused on data inclusion and consideration for system dynamics modeling procedures. The authors described the possibilities for expanding the conventional system dynamics methodology of conceptualizing and formalizing models using complementary statistical methods and data collection.

Glaiel et al. [81] described in their work an investigation on how dynamics in Agile software development differ from conventional waterfall approaches. The authors built a System Dynamics model for software projects for transparency into model structure and parameters in this context. Their Agile Project Dynamics model addresses several agile focuses as separated components (e.g., user story/feature-driven and soft team factors), allowing a simulation-based exploration of positive and negative effects through adjustable development and management practices. Moreover, the model supports dashboards that facilitate investigations for managers without system dynamics expertise.

3.3   Feedback and Information Visualization

Agile software development is known for its intended short development cycles, thus quickly outdated progress information. Project management systems (e.g., Jira, Trello, and Asana) help teams and managers keep track of daily development tasks and progress. The availability of feedback for teams (from customers or managers) and other improvement-relevant information is crucial but often taken for granted. Only a few studies concern computer-aided feedback for teams in agile projects, which was the central focus of this thesis.

Vetro et al. [233] studied the effects of fast feedback in agile software development. The authors' systematic mechanisms for collecting additional feedback allow extracting knowledge supplementary to the informal insights in retrospective meetings. They propose data analyses combined with automatic mechanisms for capturing stakeholders' feedback and improving the analyses' precision and applicability. The authors observed the impact of the additional feedback mechanisms by collecting information from agile development teams and identified the resulting implications for the quality and knowledge gained for the next sprints (e.g., estimation of user stories with reduced human effort).

Lehtonen et al. [153] disclosed the key to improvements in the agile context is information visualization. The authors found that agile methodology emphasizes team reflection to make problems visible and learn from the past. While software projects quickly produce many data, they can be mined and visualized for process improvement purposes. Teams can increase their awareness of the causes of particular problems because of the human mind's capability to interpret visual representations better than purely textual information. The authors conducted an action research approach in the industry, which resulted in visualization support for data stored in an existing issue management system. Their research revealed that the issue data provided more insightful knowledge through complementary visualization techniques, allowing them to identify more manageable problems. The authors concluded that a visual approach could successfully point out the issues and inferences in issue handling, which would have been hard to find otherwise.

Jermakovics et al. [110] studied the benefits of analyzing and visualizing developer networks. The authors characterized different team members' interactions based on data from repositories to identify and highlight workloads and internal development structures. The authors' research focus was a central aspect of this work that has led to socio-technical sprint dependency visualizations expressed through force-directed network graphs.

Hevner et al. [93] described information systems research as the process to provide an understanding of a problem domain and its solution. The authors used a behavioral science paradigm to seek, develop, and verify theories explaining or predicting human or organizational behavior. The design-science paradigm was used to explore and extend human and organizational capabilities by creating new and innovative artifacts [92].

Chapter 4

Concept for Capturing Socio-Technical Aspects in Agile Projects

Agile software development (ASD) is an activity that relies on people whose behaviors are directed by sociological and technical factors [111, 219, 238]. Patterns of interactions among team members and the resulting development performance determine the positive and negative team dynamics [62, 122]. However, characterizing the team dynamics in ongoing software projects can be challenging due to the limited availability of sociological data and dynamic changes of complex socio-technical dependencies within the sprints [134]. In this context, this work presents two concepts related to the systematic capture of socio-technical aspects in agile software projects and the support of computer-aided sprint feedback.

This chapter describes the first concept, which concerns the systematic capture of quantifiable data on team dynamics optimized for agile software projects. The systematic collection of objective and subjective data that characterizes team dynamics is relevant for the second concept presented in Chapter 5, which pertains to computer-aided sprint feedback (e.g., the knowledge and data foundation for descriptive, predictive, and exploratory models). The computer-aided sprint feedback conceptualization involved a Goal Question Metric (GQM) process, which disclosed relevant socio-technical metrics for understanding team dynamics in ASD [17]. Moreover, the GQM process subsequently led to several measurement methods that systematically capture the metrics through objective data tracking in project management systems and subjective opinion polls during the software process based on team reflections. However, it is impossible to objectively consider all aspects of a projec [101].

The concept enables project members to consistently collect socio-technical data concerning the interrelations between individual team's behavior and objectively measurable development progress during sprints with maximal autonomy. At the same time, the involved metrics and the associated measurement methods rely on scientific foundations. The concept has resulted in a series of refinements for operationalized data capture and the characterization of team dynamics [134].

4.1   Relevance of Systematic Measurement Methods

Organizational cultures impact how individuals perform tasks and live their day-to-day work lives [144]. Therefore, it is crucial to understand teams' behavior and the related dynamics, which occupy a significant role in the effectiveness and improvement of individuals' and teams' daily operations [121]. If a team is not aware of underlying team dynamics, improvements are challenging to place [103]. In this context, understanding team dynamics in software projects begins with systematic measurements of the socio-technical metrics that characterize development behavior during the software process. From an organization's perspective, it promotes building an insightful and knowledgeable foundation based on the captured data, thus impacting factors related to the projects' progress and drawbacks to enable future improvements.

Technologically advanced tools that support the systematic measurement of soft factors in development teams are rare. The existing ones often lack a scientific foundation or the lightweight methods necessary to capture the subjective team aspects needed for holistic characterizations of team dynamics [122, 219]. As a result, the retrospective identification and interpretation of strengths and weaknesses in teams is often only superficially performed in projects [17, 122, 138]. Researchers have previously conducted several empirical studies to understand team behavior in industrial and academic software projects, but the research findings often refer to limited case implications without generalizability [62, 126, 179].

Capturing team dynamics in a real-word setting can be a challenging task due to the external interference of study-related effects in otherwise natural development environments [121]. What is referred to as the Hawthorne effect can occur in group-based observational studies, when participants alter their natural behavior because they know that they are participating in a research study and are under observation. This alteration can lead to an incorrect assessment, such as mocked behavior or development effectiveness [147]. Possible explanations for the Hawthorne effect include the impact of feedback and motivation on participants. Feedback on performance may improve developer skills when a study provides this feedback for the first time [182]. Therefore, capturing behavioral aspects in a full-situated development environment requires measurement methods beyond standard self-reports for quantitative observations.

4.2   Human Factors Associated with Team Behavior

Software development builds on successful teams, which rely, among other factors, on human factors and the related behavior of developers [53, 111, 172, 207]. In software engineering, focus on the soft factors in development teams and their effects within projects has increased over the years [179, 198, 201]. The performance and commitment of individuals, their adduced moods, and the related social interactions within teams are central factors that promote an understanding of team dynamics in software projects as a whole [62, 121]. In the case of reoccurring events or situations, the detection and interpretation of patterns connected to behavioral factors and development outcomes can explain performance changes in projects over time [170, 173]. Moreover, team performance is an essential element for project success and is exceptionally vital for team achievements in highly volatile agile development environments [70].

Understanding the underlying effects on development performance is the basis for guiding the formation and maintenance of high-performing teams and informing performance improvement initiatives and environmental conditions (e.g., mutual trust among members, communication effectiveness, group thinking, diversity and competencies of individuals, and leadership support) [70, 219] The combination of objective progress measures and sociological factors enhances the characterization and interpretation reliability of team dynamics [2, 62, 70]. Team performance centers around behavioral and social sciences. At the same time teams are engaged in knowing best their performances towards matching their effort estimations with stakeholders' expectations [70].

Therefore, the availability of useful information related to human factors in projects supports many process-related activities (e.g., planning, progress tracking, software quality measurement, process problem awareness, and the motivation of people [78, 143]). A set of team behavior-related human factors is summarized in Table 4.1. They resulted from previous literature reviews on social factors in software engineering and are relevant to characterize different aspects of team behavior [25, 67, 100, 111, 143, 198]. The subsequent sections cover the systematic identification of related metrics and measurement methods in the context of ASD.

Table 4.1: Set of Relevant Human Factors from Software Engineering Literature

Category	Literature Reference
Mood	[96, 152, 205, 207, 236, 244]
Communication	[62, 145, 173, 207, 218]
Satisfaction	[56, 83, 132, 180, 231]

Mood in teams relates to the happy or unhappy feelings of individuals due to positive and negative experiences collected during the software development process [60, 236, 244]. Teams with an overall positive mood perform better (e.g., communication activities), promoting higher customer satisfaction [125, 159]. Team members' well-being is a crucial factor relating to projects and teams' environmental working conditions. The team mood in a project provides valuable insights into the teams' sociopsychological functioning (e.g., the handling of social conflicts or task-related problems, self-esteem in the group, and changes in the commitment or group thinking) [23, 99, 205, 238]. Chapter 2 introduced the fundamentals from social and behavioral psychology with well-evaluated methods for systematically capturing teams' moods, whereas the knowledge-supporting goals from practitioner perspectives must be identified.

Communication in teams is crucial for project success [2, 34, 62, 138]. Social interactions among team members without social barriers promote activities such as cooperating, sharing experiences, discussing problems jointly, and determining improvements as a team [115, 173]. Direct communication in software projects subserves a relationship of trust and harmony [219]. Moreover, it promotes developers' social and interpersonal skills [150]. Communication gaps or reduced external autonomy are significant barriers for self-organizing teams with implications for team performance [3, 32, 33, 173]. The observation and interpretation of communication structures over time through social network analysis supports the understanding of information flows in projects and the related social interactions in teams [90, 206, 242].

Satisfaction is deeply anchored to the perceived happiness in teams, also by the project leader, and the customer [10, 239]. Satisfaction in software projects is often perceived differently based on the underlying definition and understanding of project success as they pertain to the individual perspectives and expectations [125]. Software project success and failure often relate to the stakeholders' satisfaction with and acceptance of the realized software features, as well as obtaining the management's desired economic goals. Therefore, stakeholders' satisfaction is an early indicator of the perceived performance in a project [180]. Project leader satisfaction often refers to the teams' well-doing concerning the functional compliance between the development progress and effective teamwork needed to fulfill the desired product and economic goals [173, 244]. Frequent feedback concerning the project leader and stakeholders' satisfaction promotes an early success reference and has been reported as important to support agile teams in sprints [132, 138].

4.2.1 Activity-Related Perspective on Team Performance

Chapter 2 provided a general overview of the socio-technical factors and behaviors in agile software projects (e.g., team communication and interactions). These were prerequisites for identifying relevant metrics for computer-aided feedback on the team dynamics during the development processes. Moreover, it is also necessary to consider different development activities to accomplish goal-focused feedback with practical support concerning insights on complementary team dynamics. This socio-technical data capture concept consider three main development activities that rely on the widely applied agile method Scrum, related agile values from the agile manifesto and experiences collected in industrial software projects: *Review* (past performance), *Plan* (future performance), and *Implementation* (Actual Performance) [22, 24, 128, 142, 164]. Figure 4.1 depicts the activity-related perspective on team performance in agile practices.

The three activities in a sprint cycle depict a differentiated need for sprint feedback (e.g., insights about the actual weeks' development performance during the implementation, a summary of related behavioral changes across all previous weeks in sprint reviews, and the implicated trends relevant for sprint planning).

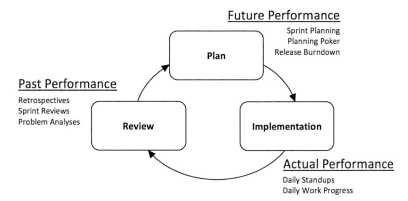

Figure 4.1: Activity-Related Perspective on Team Performance in Scrum

As performance-related information is quickly outdated, feedback on team dynamics must be made available during the particular activities without delays to be of practical value. These environmental constraints impact not only the identification of socio-technical metrics and measurements but also the capture intervals. Section 4.3 covers identifying goals, assessable questions, and quantitative metrics using GQM to improve the understanding of team dynamics in ASD.

4.3   Team Dynamics Feedback: Goal, Questions and Metrics

The starting problem of understanding team dynamics in agile software development (ASD) is to understand "how do you decide what you need to know to achieve your goals?" [17]. The GQM method supports defining a set of goal-focused questions, each to characterized a specific team dynamics aspect that, in the end, can be quantitatively answered through metrics applicable in ASD [17].

The goals, questions, and metrics defined during the GQM definition activity originate from interdisciplinary foundations (see Chapter 2), related work (see Chapter 3), and personal observations in academic and industrial software projects [128, 129, 135, 139]. Rather than developing new metrics, the GQM method helps defining concrete metrics and practical assessment methods based on already well-evaluated foundations and the constraints of the ASD domain.

In this work, the perspectives of different project roles, agile activities, and interdisciplinary foundations are covered in the definition process to establish a socio-technical data foundation needed for the data capture concept and automated processing of the computer-aided sprint feedback concept (see Chapter 5). The result is a GQM model that involves the definition of concrete goals helping to understand team dynamics better in ASD. Figure 4.2 shows the included ASD information assets associations relevant for the *GQM-Definition* activity.

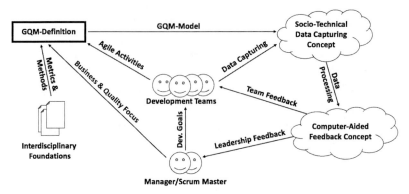

Figure 4.2: GQM-Relevant Information Assets and Associations, based on [17]

4.3.1   Conceptual Goals for Understanding Team Dynamics

GQM intends to identify measurement goals with practical relevance (e.g., to understand team dynamics in agile project environments) [17]. The first step of the GQM definition targets four conceptual understanding goals related to sociological and behavioral aspects in agile development teams (see Table 4.1) and objectively assessable performances measures. The goals rely not only on the relevance of the metrics in explaining team dynamics but also on the expected practicability in ASD. The goals are divided into subgoals defined by more concrete aspects interrelated with team dynamics. The refinements include interdisciplinary foundations from software engineering and social and organizational psychology, as well as practitioner expertise about the information needed for better understanding team dynamics [62, 122, 179, 207].

The goals contain unique properties concerning at least two or more underlying factors to be considered by the different perspectives of the corresponding project roles (i.e., the development team, manager, or Scrum Master). All goals are abstracted using a template scheme for disclosing their quality focuses (see Appendix. C.1). The abstraction of the goals is an essential activity concerning the definition of subsequent operational questions, which are used to assess whether the goals have been fulfilled. Figure 4.3 shows the conceptual goals and subgoals in this work.

4.3.2   Operational Questions for Answering Goals

The second step of the GQM definition activity resulted in question sets describing the abstract goals in a quantifiable way concerning their quality focus. The questions are used to determine whether the goals' purposes is fulfilled through adequate, quantitative measurement methods. In this work, abstraction sheets were applied to concretize the measurement goals systematically. The question sets derived from these sheets were a prerequisite to each goal's quantitative assessment and thus an improved understanding of team dynamics in ASD.

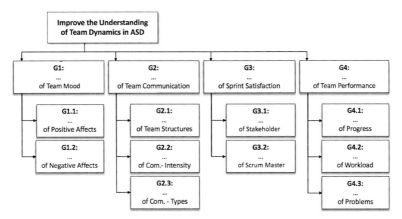

Figure 4.3: Abstract Goals for Understanding Team Dynamics in ASD

For the definition of the goal-assessing questions, the abstraction sheets cover a header, four different quadrants, and a footer [17]. Each quadrant is associated with a goal-related quality aspect. Figure 4.4 shows the applied question definition process using abstraction sheets for G1 (team mood).

⇒ *The header* summarize properties of the targeted goal template G1. In this example, the team mood is the study object. The purpose is to understand the quality focus of the perceived mood levels from the development teams' perspective within the environmental context of ASD.

⇒ *The quality focus* describes concrete measurable properties of the targeted object. In this case, the quality focus is defined towards understanding the mood changes in teams by knowing the members' perceived positive and negative feelings in sprints.

⇒ *The baseline hypothesis* describes the normal state of the targeted quality focus. ASD builds on people-centered practices. However, team mood is often only superficially reflected in sprint retrospectives or inconsistently captured during the development process. Thus, the understanding of mood changes over time and is limited in quality and validity.

⇒ *The variation factors* covers the presumed influencing factors for the described quality focus—consistent scales for systematic mood reflections in teams can depict positive and negative perceptions.

⇒ *The Impact on Baseline Hypothesis* describes the expected change caused by the variation factors. Consistently capturing and tracking team moods enables the changing emotional states among the members and over time to be compared. It provides an equal interpretation base for the mood in teams.

⇒ *The Feedback* provides an optional remark relevant to the targeted quality focus whose influences are under examination. It concerns the interrelation of productivity and team mood.

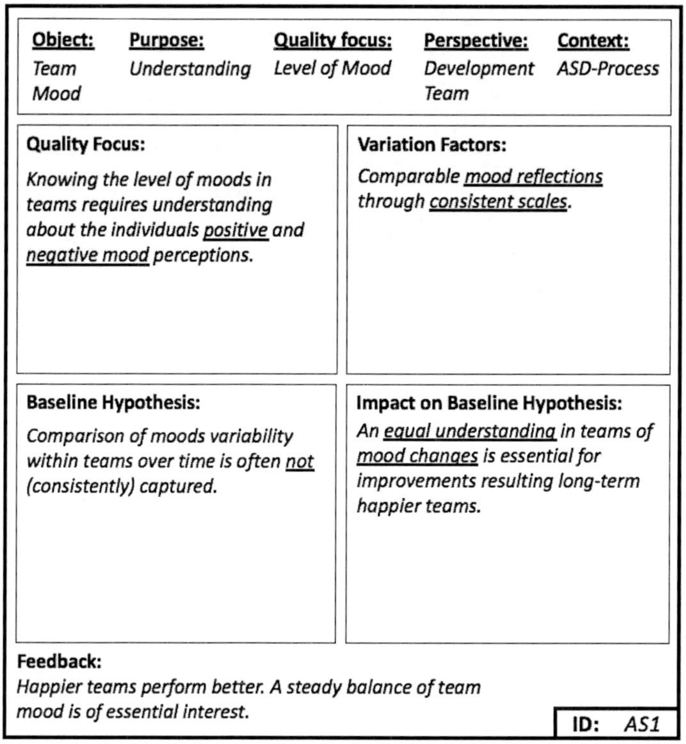

Object:	Purpose:	Quality focus:	Perspective:	Context:
Team Mood	Understanding	Level of Mood	Development Team	ASD-Process

Quality Focus:	Variation Factors:
Knowing the level of moods in teams requires understanding about the individuals _positive_ and _negative mood_ perceptions.	Comparable _mood reflections_ through _consistent scales_.

Baseline Hypothesis:	Impact on Baseline Hypothesis:
Comparison of moods variability within teams over time is often _not_ (consistently) captured.	An _equal understanding_ in teams of _mood changes_ is essential for improvements resulting long-term happier teams.

Feedback:
Happier teams perform better. A steady balance of team mood is of essential interest.

ID: AS1

Figure 4.4: Abstraction Sheet for "G1", according to [17]

The questions set listed in Table 4.2 is derived from the quality aspects of the abstraction sheet $AS1$ so that the targeted $G1$ can be described as quantitatively as possible through adequate measurement methods.

Table 4.2: GQM Definition Step 2: Example Defined Question Set for $G1$

Question	Goal-Question-ID
How intense is the perceived negative mood?	*G1.Q1*
How intense is the perceived positive mood?	*G1.Q2*
How is the level of mood in teams?	*G1.Q3*

The goal-related questions were directly discussed with practitioners for relevance. The abstraction sheets $AS2$ to $AS5$ for the other conceptual goals are listed in Appendix C.2) and the derived question sets for $G2$ to $G5$ are listed in Appendix C.3. The question definition is equally processed for all the abstract goals and an essential step before determining the relevant metrics and practical measurement methods for assessing the questions.

4.3.3   Quantitative Metrics for Assessing Questions

The third step of the GQM definition activity results in a set of formally defined objective and subjective metrics that allow the previously derived goal-based questions to be assessed (see Appendix C.3). The quantitative assessments determine whether each goal's purpose has been satisfied (e.g., improved understanding concerning goals' socio-technical focus). The GQM process does not explicitly differ depending on whether existing metrics are identified and used or a metric must be newly defined (e.g., a combination of existing or new measurements).

In practice, most ASD metrics used during the development process are related to objective measurements (e.g., development progress, team productivity, and code quality). Subjective measures are rarely applied, as the overview in Appendix A.1 depicts. For completeness reasons, the terms "objective measure" and "subjective measure" in this work correspond to the following definitions:

Definition 1:   An **objective measure** is a metric (i.e., a mapping from the object to be measured to a number). The mapping can be presented as a concrete mathematical formula, allowing for an objective and reproducible result.

Definition 2:   A **subjective measure** is the object measured based on the perspective from which it is taken. Examples include perceived communication intensity, the level of stakeholder satisfaction, or the mood.

Consequently, the perspectives of different project members are involved during the metrics identification to ensure that a goal's quality focus does not result in different interpretation scopes based on the member role (e.g., perceived success) [125]. The resulting number of subjective measurements dominates the set of metrics pertaining to the abstract goals of understanding the team dynamics in ASD. Chapter 2 and 3 indicated several relevant metrics interrelating to team behavior and thus valuable for understanding team dynamics in ASD. No entirely new metrics are established to answer the goal-related questions quantitatively. Each question is assessable using at least one or more relevant metrics from already well-evaluated research foundations.

The following process is presented for deriving metrics using the example of questions related to $G1$ (understanding the mood in teams). Knowing the team moods in ongoing projects promotes awareness and indicates functional development environments [83, 125, 159]. Awareness of mood changes over time enables teams to recognize and react to early stages of social conflicts or task- and process-related problems [23, 238]. In industrial agile software projects, capturing the team's mood is often accomplished unsystematically (e.g., using stickers anonymously placed on a board with positive, neutral, or negative fields corresponding to members' mood at the end of a workday). While an offline method is lightweight and generally provides initial indicators of a team's overall mood status, it is inefficient for a systematic comparison with other socio-technical aspects.

Moreover, the differentiation between solely positive, neutral, or negative states is less expressive and insightful than established measures from psychology (e.g., the PANAS) [236]. A team with a strongly positive mood tends to be happier, more active, and alert, while an intense negative mood tends to cause angrier, more nervous, or feared member [96, 244].

Concerning the quality focus of $G1$ and the subsequent derived assessment questions in the practical context of ASD, the lightweight mood-affect measures from the short form of PANAS are used to assess the positive and negative team moods quantitatively [207, 225]. The applied short form encompassed 12 mood descriptors, each quantifiable through a five-point rating interval to systematically capture the subjective perception of members' individual mood aspects (see Section 2.1.5). For example, answers for the self-rated mood question "To what extent did you feel nervous during the last week?" range from 1-never to 5-always (see Table 2.2).

In practice, the short form of PANAS has been widely accepted for its 12 self-rated mood characteristics instead of one because it allows for nuanced mood characteristics rather than the differentiation between positive and negative [225]. From a psychology perspective, the combination of these mood aspects allow characterizing a broad spectrum of complex emotions with relevance for teams, and their resulting project performance [196, 224]. The median of all the weekly captured developer perceptions is derived to describe the overall team moods more objectively [244]. Moreover, aggregating moods as a team value granted anonymity for members who feared exposing their personal feelings. The definitions of the metrics for positive and negative mood are described as follows:

$$\text{negativeMood}_{i,j}(t) \in \{1,\ldots,5\}, \text{where } t = \text{Projectweek} \tag{4.1}$$

$$i \in \{\text{tempted,confused,worried,angsty,nervous,angry}\} \tag{4.2}$$

$$j \in \{\text{Developer}_1,\ldots,\text{Developer}_{Teamsize}\} \tag{4.3}$$

$$\text{negMoodAffect}(t) = \text{median}_{i,j}(\text{negativeMood}_{i,j}(t)) \tag{4.4}$$

$$\text{positiveMood}_{k,j}(t) \in \{1,\ldots,5\}, \text{where } t = \text{Projectweek} \tag{4.5}$$

$$k \in \{\text{active,interested,happy,strong,encouraged,awake}\} \tag{4.6}$$

$$j \in \{\text{Developer}_1,\ldots,\text{Developer}_{Teamsize}\} \tag{4.7}$$

$$\text{posMoodAffect}(t) = \text{median}_{k,j}(\text{positiveMood}_{k,j}(t)) \tag{4.8}$$

The formal definitions of all the metrics associated with assessing the goal-related questions for $G2$ to G are attached to Appendix C.4. All the GQM associations are concluded in a GQM model (see Appendix C.5), which is the final step of the GQM definition activity that summarizes the defined goals' quality aspects (i.e., the specified purpose of the measurement), derived assessment questions (i.e., the defined quality focus of the goals), and quantitative metrics (i.e., the identified measures for the quantitative answering of questions). The model discloses the hierarchical mappings between the specific goals, derived questions, and defined metrics and forms the foundation for socio-technical data capturing and computer-aided feedback concepts. However, several of the metrics summarized in the GQM model involved subjective measures based on project member perceptions and introspective self-assessments concerning the sociological team aspects (e.g., the perceived team communication intensity). In practice, these subjective measures are often challenging to systematically capture without causing study-like constraints or significant additional effort for the teams [122, 232].

4.4   Measurement Methods for Aspects of Team Dynamics

The following methods concern the systematic capture of the identified socio-technical metrics, thus assessing each goal's satisfaction. Systematic capture of socio-technical aspects in ASD emphasizes practical information management systems (e.g., how agile teams share and access development information during the sprints). Observations and interviews conducted on industrial software projects provided knowledge about teams' development processes [22, 128].

Jira is a well-established project management system to monitor agile software projects noted during the cooperation with industrial research partners. As a result of these onsite experiences and related work, measurement methods were conceptualized for the data capture concept and prototypically developed for the project management system Jira [153, 232, 233]. The prototypical realization of the socio-technical data capture and computer-aided feedback concepts for Jira is practical. To evaluate the usability and utility of the concepts without study biases or framework limitations, the best practices of agile development teams from the industry are essential. In this context, a persistent constraint in academic and industrial software projects concerns the processing of private information from individuals and teams. Consequently, these concerns are identified and thematized in the following subsection before introducing the socio-technical measurement methods.

4.4.1   Constraints on Capturing and Processing Personal Data

The identification and conceptualization of practical methods that support the systematic capture of socio-technical data were derived from observations in academic and industrial software projects. The few objective measures of $G4$ (understanding team performances) in the GQM model (e.g., the progress and workload in sprints) did not pose a specific challenge for systematic tracking due to the native availability in standard project management systems. However, the systematic data capture of metrics related to subjective measures is more challenging, also due to restrictions regarding the information privacy of team members (e.g., mood or communication behaviors).

Information Privacy in the digital age is a sensitive topic considering the faster collection and processing of larger volumes of personal data [26, 160]. Advances in information technology have globally raised concerns about information privacy, and its impact, especially for data that can be obtained, aggregated, or analyzed without individuals' awareness [26, 160]. Information Privacy concerns the traceability of personal information, which endangers individuals' anonymity and privacy. Particularly in Germany, where most of this work's conceptual research was done, managers were highly cautious about maintaining employees' information privacy while trying to understand better the team dynamics in sprints [128]. Moreover, the observations allowed recognizing teams' identities, structures, internal member competitions, and individual privacy concerns. In this context, the following information privacy criteria, which correspond to Clarke's [46] four dimensions of privacy, are considered by this work in the data capture and feedback processing concept:

- Privacy of a Person

- Personal Behavior Privacy

- Personal Communication Privacy

- Personal Data Privacy

Feedback culture and knowledge sharing in organizations are often contingent on fears about the loss of autonomy, which would hinder the team's ability to accomplish their tasks if the team grants access to their data [162]. Nevertheless, feedback promotes teamwork, reduces social barriers, and enables process improvements [99, 115, 170, 239]. However, not every team member places personal benefits below the group's, especially when the long-term benefits are not foreseeable. A prior study disclosed that people's attitudes could change based on being told that their data is processed and managed using fair information practices [51]. Such attitudes include sensitivity to sharing or loss of information or willingness to share personal information [51]. When their concerns for information privacy or fear of losing autonomy are sufficiently mitigated, team members will support providing personal information. Otherwise, they will protect their positions [29]. An international survey study with researchers and practitioners revealed that most of the respondents shared an open mentality towards providing their colleagues with personal feedback about their own experiences and perceptions collected in sprints, as long others' attitudes followed a similar mindset [138].

Management Systems are the information memory of agile software projects, and they host valuable data about teams' daily development progress. In widely-used project management systems, such as Jira or Trello, all process-relevant data (e.g., the estimated story points, reported bugs, productivity) is systematically collected and made available for every project [155]. Sociological aspects interrelated with team performance are natively not tracked or systematically captured during agile development activities. The shared problem is the lack of systematic methods and tools or the time required by teams to improvise practices to gather insights on team aspects. Teams often return to simple task-based status boards or burndown charts that allow them to track the completed and remaining work and thus the likelihood of achieving the sprint goals [128]. Continuous process improvements based on systematic analyses are inefficient for manual information processing (e.g., offline elicited moods), especially when sprint reviews are time-limited.

4.4.2   Interval-Based Measurements in Agile Software Development

The information constraints are considered and aligned for agile development activities in this concept. As motivated by related work, fast feedback on team dynamics has been accomplished through the embedded capturing and processing of socio-technical data within an existing project management system [139, 233]. The socio-technical data capturing concept is realized through an add-on for Jira and complements teams' standard agile development environments by an interval-based measurement method for socio-technical data. The focus of the measurement methods relies on subjective data capturing of the team members' perceptions. Figure 4.5 shows the interval-based data capture concept based on objectively tracked team performance data, subjective sociological data from the team, and the sprint satisfaction of the customer and Scrum Master.

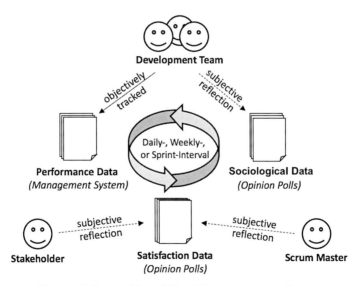

Figure 4.5: Interval-based Data Measurements in Sprints

The capturing methods for the subjective data are realized using opinion polls (customizable survey template). They are made available through an add-on service layer for existing project management systems (e.g., prototypically for Jira in this work) [174]. The service layer automatically administrates different opinion polls by the project roles in predefined intervals (e.g., weekly team reflections and customer reflections at the end of sprints) and manual invitations anytime based on a particular event. Push notifications invite team members to participate in a web-based questionnaire within Jira. External customers and Scrum Master receive an email invitation that directs to the project-related sprint satisfaction survey in Jira.

Performance Data: The development progress in agile teams is tracked objectively using quantitative measurements concerning the development task (issue) properties (e.g., transition times and workflow states). In this work, the tracked tasks types included stories, subtasks, and bugs, each observed corresponding to their respective estimated complexity based on story points and the development workflow states (e.g., to do, in progress, and done). In this context, conventional performance metrics, as introduced in Chapter 2 are natively available (e.g., the velocity or estimation error) to systematically answer the defined questions related to $G4$ (understanding team performance). Each status change of tasks indicates progress.

With this, the number of day-by-day completed tasks and story points and the scheduled and actual number of developer capacities in a sprint is objective progress measures over time. Ideally, the number of incomplete tasks and story points should decrease over time to complete all the scheduled tasks following the initial sprint estimations. The performance data is therefore tracked daily and compiled in aggregated performance summaries for specific weeks and sprints. What information a development task contains in detail depends on the applied project management systems. However, development tasks in Jira commonly provide supplementary, less apparent details that support further objective characterizations (e.g., the team interaction structures during sprints based on task-related comments, support requests, and task forwarding). Supplementary task details that are covered by Jira are shown in Appendix A.2.

Sociological Data: Understanding the team dynamics in ASD through computer-aided sprint feedback requires the systematic capture of sociological team data in sprints. This concept builds on the advantages of technological support (i.e., web-based survey that enables the automated processing of team member responses for fast sprint feedback in commonly time-sensitive development iterations). The digital survey captures the sociological data from the teams in a predefined interval, based on well-evaluated measures and measurement methods defined in $G1$ (understanding team mood) and $G2$ (understanding team communication)in this concept's GQM model (e.g., using PANAS and SNA).

However, the questions and self-assessment answer options concerning the metrics defined in $G2$ are reduced in complexity and manual effort based on previous research described by Schneider et al. [206, 207]. As described by the authors, paper-based opinion polls are a suitable and well-established form to collect participants' feedback for a case study. The authors' used weekly paper questionnaires to elicit the communication network based on who-to-whom statements and communication intensity ratings for multiple media channels. Nevertheless, for a fast sprint feedback routine in (possibly multiple) agile software projects, such an offline method is impractical due to the handling time, manual effort, data processing, and analysis.

Figure 4.6 shows an abstract example of the digital survey structure that resulted from a sequence of iterative improvements in academic software projects (cf. [139, 207]). The digital survey is iteratively improved, along with the transition from a manual paper-based process to an externalized mobile application support to an adaptive web-based method. In contrast to initial paper-based studies, digital capturing and processing optimized the automated analyses of team behaviors.

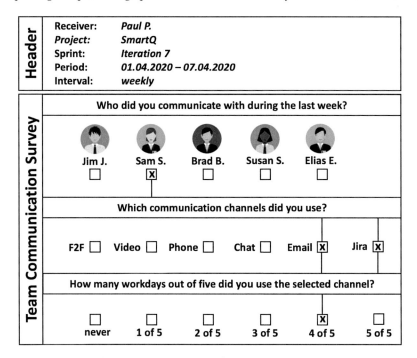

Figure 4.6: Subjective Measurement of Team Communication in Sprints

Moreover, the digital survey enable the integration of functional advantages that reduce time and effort (e.g., follow-up questions about communication networks are only asked depending on the respondents' answers and canceled if not relevant) [134]. The survey interval is adjustable, corresponding to the participation preference in teams. The sociological team survey about the subjective measurements of $G1$ and $G2$ is attached in Appendix D.3.

Satisfaction Data: Project success is a highly subjective aspect that often differs between roles-related perspective (e.g., that of the stakeholder, manager, team leader, or team members) [125, 190]. Feedback is essential for a team to improve and determine whether the completed work in sprints will result in satisfaction or dissatisfaction across the roles [180, 239]. However, the objectively measured performance metrics from $G4$ (understanding team performance) and the subjective team behavior measurements from $G1$ and $G2$ only considered the objective and subjective data provided by teams. The digital survey for the systematic measurement of the stakeholder (or product owner) and Scrum Master or project manager's sprint satisfaction is included in this concept to provide teams supplementary feedback from external perspectives regarding their performance during the previous sprint. The external input allows comparing satisfaction changes over time together with the socio-technical aspects captured in sprints. The survey interval was event-triggered so that both roles automatically received an email invitation to participate in the sprint feedback survey.

The understanding goals concerning external sprint satisfactions were specified in $G3$ (see Appendix C.5). In this context, Kirkman's and Rosen's scale [117] described in Section 2.1.6 is a well-evaluated method that was used for subjectively measuring team performance in addition to the objectively tracked sprint performance data related to $G4$. The initial question set and rating options were inherited in this digital version of the sprint satisfaction survey. The realized team performance survey covered four questions concerning the team's performance and four questions concerning the development performance (see Section 2.1.6). These questions allowed for a detailed characterization of relevant performance aspects based on individuals' perceptions. Correspondingly, the digital version also asked the participants to rate their overall satisfaction during the last sprint for each accomplished team performance and development performance using the five-point interval scale. Appendix D.3 includes the complete survey.

The realized survey administration support for Jira automatically tracks responses and reminds participants of a pending survey [174]. Moreover, this digital survey's technological advantage promotes short communication path in sprints for faster feedback, especially when the stakeholders' physical attendance is not possible. The stakeholder and Scrum Master's dual participation increases the likelihood of detecting significant differences in the perceived sprint performance, allowing for considerations and adjustments in next sprints. The computer-aided sprint feedback concept in Chapter 5 covers the consecutive processing methods for analyzing socio-technical aspects in sprints complementary to the data capture concept.

4.5 Technological Feasibility of the Data Capture Concept

This concept aims to specify relevant goals concerning understanding team dynamics in ASD and derive metrics with practical measurement methods. The concept is prototypically realized for Jira with the particular focus of a consecutive evaluation of the computer-aided sprint feedback concept (refer to Chapter 5). Therefore, the initial approach of the technological feasibility described in this section only comprises a series of practical approaches to the measurement methods applied in academic software projects. Concerning the GQM activity covered in this concept chapter, the explicit conformance of the GQM goals based on the accomplished computer-aided sprint feedback's usability and utility was assessed in agile projects from the industry (see Chapter 6).

However, the general technological feasibility of the socio-technical data capture methods was approached using observational studies during the years 2019 and

2020 involving two cohorts with 130 and 147 undergraduate students each, in a total of 32 software projects and duration of 15 weeks [134, 139]. Figure 4.7 depicts
the observed student software projects' data capture process. Participation was optional and depended on the teams' motivation and desire for advanced sprint feedback. The observations showed the concept's functionality in practice and helped
improve the capture methods' technological maturity (e.g., reducing the survey
complexity and enhancing the clarity of the descriptions) [134, 136, 139].

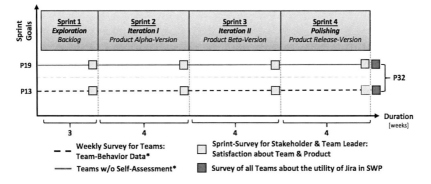

Figure 4.7: Data Capturing Concept Applied in Student Software Projects

The systematically captured socio-technical data in the student development teams
and the satisfaction reflections of the project leaders and customers at the end of
the sprints allow building a data-driven knowledge base. The latter is described
in Chapter 5 and was essential for establishing the computer-aided sprint feedback
concept in this thesis (e.g., the methods for automating sprint analyses and models) [134, 136, 139]. The consecutive processing of all the data for advanced sprint
feedback on the team dynamics in ASD is described in Chapter 5.

Chapter 5

Concept for Computer-Aided Sprint Feedback

Rapidly changing technologies and project challenges (e.g., high product expectations while maintaining economic profits) have increased the relevance of efficient information management in organizations [167]. In practice, most software companies utilize information systems (IS) in their projects to support the development teams with useful, process-related information (e.g., progress reports, the status of development tasks, and workflows) [55, 94]. While important, systematic feedback also on the sociological aspects in software projects is barely provided.

The computer-aided feedback concept described in this work aims to provide fast feedback that supports the team understanding of arisen team dynamics in sprints. Thus it considers the socio-technical factors in project environments essential for interpreting team behaviors and related development performances. Chapter 2 and 4 previously discussed socio-technical aspects with relevance for understanding the team behavior in agile software development. The set of identified metrics and measurement methods enable the data foundation for a systematic characterization of team behavior in sprints. The emerged socio-technical data capture concept forms the basement for this concept about computer-aided sprint feedback on the team dynamics in agile software projects.

The concept extends the progress information in agile software projects through systematically processed socio-technical data for descriptive, predictive, and exploratory feedback on the team behaviors in and across sprints. The analytical methods of the three feedback assets differ regarding their processing purpose. The descriptive asset builds on statistical analyzes to describe team behavior courses and anomalies in the past. In contrast, the predictive asset uses the past data as a foundation for machine learning models that subsequently compute prospects about future team behavior. In comparison, the third asset relies on exploratory sprint dependency models and a system dynamics model for sprint simulations. The computer-aided sprint feedback processing follows the principles for a simplified knowledge discovery process in databases (see Section 2.2.2).

A supplementary visualization concept for the feedback assets fosters the cognitive understanding in teams concerning the performance-related development behavior in and across sprints. The applied techniques for visualizing the sprint feedback evolved during design science and action research based on interdisciplinary foundations, related work, and user feedback. Figure 5.1 summarizes the processing layers of the computer-aided sprint feedback concept built upon the socio-technical data capture concept described in Chapter 4.

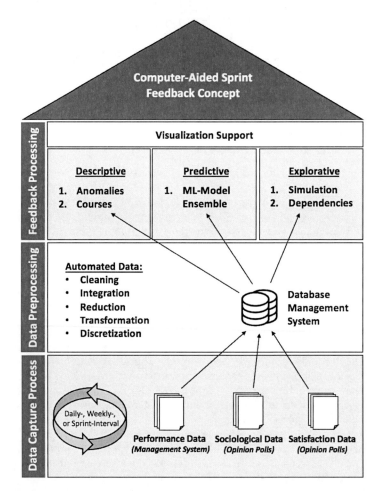

Figure 5.1: Process Layer of the Computer-Aided Sprint Feedback Concept

The technological feasibility of the concept's prototypically realized ProDynamics plugin for Jira was approached during several observational studies involving 32 student software projects [134, 136, 139]. Chapter 6 describes the additional assessment of ProDynamics sprint feedback support in practice concerning its utilization and utility in two industrial software projects.

5.1 Preprocessing of Socio-Technical Data

Design science research stands out because it incorporates data analyzing techniques, model-based methods, and goal-related measures that together enable the transformation of environmental data about human factors, development processes, and other project characteristics into knowledge-promoting teams and management information [93, 165].

As often is the case in data-driven approaches, the raw form of data can contain undesirable entries. These require preprocessing to meet the input requirements of descriptive sprint analyses and predictive and explorative modeling methods (e.g., removal of "null" values or irrelevant attributes). The collected socio-technical data is stored in several data tables, extending an existing database and table structures of a project management system (e.g., Jira in this work). Newly captured sprint data have to undergo data preprocessing routines for handling data-related issues. Such routines can include missing values, cleaning up data errors or inconsistencies, transforming data attributes, or normalizing differing data ranges if needed for statistical operations [148, 221]. Besides using a central database, storing the preprocessed socio-technical data is relevant for the subsequently processed feedback to retain data timeliness and quality, especially for agile development, where data often alters the moment it is collected. The following subsections describe the socio-technical data characteristics and automated preprocessing routines for the consecutive sprint feedback processing.

5.1.1 Data Structure and Attribute Types

The socio-technical metrics described in Chapter 4 encompasses different data types corresponding to several objective and subjective measurement methods (e.g., interval-scaled data from external opinion polls and continuous data from internal task-activity tracking). Newly captured socio-technical data is separated into multiple tabular forms within the database (e.g., one table stores only the opinion polls' communication data). Each row entry corresponds to a given data object (e.g., one week's communication data of a single team member). Every column describes a data object's associated attribute (e.g., member "John" communicated with member "Paul" using communication channel "email" in calendar week "10" during "4" of 5 working days). The separation of raw data corresponding to their context supports determining inconsistencies within the data due to each data attribute holds directly comparable values [221].

Additionally, all data tables include the data attributes "Project ID," "Sprint ID," and "Calendar Week" as shared composite keys to identify uniquely related socio-technical records across several tables. The calendar week results from the transaction time (epoch) at which the data has been captured. The keys store the socio-technical data in sequential order, making it possible to search or filter specific project data entries by time.

Table 5.1 lists all quantitative and qualitative data attribute types and the allowed statistical operations for processing the computer-aided sprint feedback (e.g., correlation analyses of sprint dependencies).

Table 5.1: Socio-Technical Attribute Types and Statistical Operations, cf. [221]

Types		Description	Attribute Example	Statistical Operations
Categorical (Qualitative)	Nominal	Values of nominal attributes allowing to distinguish only one object from another (=, ≠).	• ID numbers • Project roles	• mode • entropy • contingency correlation • x^2 test
	Ordinal	Values of ordinal attributes allowing to distinguish and order objects (<, ≤, >, ≥).	• Com. Channel • Com. Intensity • Com. Duration	• median • percentiles • rank correlation • run and sign tests
Numeric (Quantitative)	Interval	Values of interval attributes are meaningful, allowing to differentiate between, e.g., by units (+, -).	• PANAS • Kirkman Scale • Time & Date (Epoch)	• arithmetic mean • standard deviation • Pearson's correlation • t and F tests
	Ratio	Values of ratio attributes require the measure to be differentiable by units and the ratio is meaningful (x, /).	• Task-related counts • Team size • Workhours	• geometric mean • harmonic mean • percent • variation

The socio-technical data involves data points captured chronologically (e.g., daily, weekly, by sprints) and primarily based on ordinal and interval data attribute values (e.g., based on the rating scales in the surveys). Numeric (quantitative) data attribute types always hold number-based values, allowing arithmetic operations. However, categorical (qualitative) data attribute types contain objects or numbers without mathematical meaning (e.g., a nominal attribute type). Consequently, the statistical operation valid for a particular attribute type depends on the data value's properties [221]. Moreover, the validity of arithmetic and statistical operations based on the respective data attribute types are relevant for the preprocessing steps described in the next section (e.g., when the median is required instead of the arithmetic means due to the data attribute's properties).

The ordinal data are distinguishable and able to order (e.g., the different media richness of communication channels [123, 207]). Nevertheless, arithmetic operations are not possible, as the following example depicts. The value "5 of 5 days" of the ordinal data attribute "communication intensity" is greater than the value "1 of 5 days," but adding the value "2 of 5 days" to "3 of 5 days" is not arithmetically possible. On the other hand, interval-based data attribute values allow a simple addition. In some cases, ordinal data attributes values support permissible transformations. They can be mapped onto numerical values (e.g., from 1 = "1 of 5 days" to 5 = "5 of 5 days"), making them suitable for mathematical operations with the restriction of assumed arithmetic distances that correspond to the new numeric answers' value [221]. Concerning the previous example of communication intensity, the ordinal scale was mapped to an interval scale using an arithmetic distance of 1 (day). The socio-technical data attributes quality must be preprocessed first to obtain a clean and qualified data foundation that is readily accessible in the subsequent feedback processing.

5.1.2 Data Preprocessing as Automated Routine

The data preprocessing in this concept aims to ensure that the descriptive, predictive, and explorative sprint feedback accesses qualified data from a central database (e.g., hosted in Jira). The iterative assessment and refinement of this computer-aided sprint feedback concept led to an automatic preprocessing routine that emerged from manual data analytics methods (see Chapter 2.2.2) [35, 71, 135, 221].

Data cleaning is applied in most data mining processes to fill in missing values, smooth noisy data, or resolve data inconsistencies [221]. Consistent data quality is necessary for the involved statistical analyses and learning methods, as well as for model-driven sprint characterizations of the computer-aided feedback processing. Changing data quality can cause less reliable analyzing results or influence prediction accuracy [38]. Due to the systematic capture of socio-technical data using closed question types in surveys (see Chapter 4.4), explicit data cleaning is only needed during the conceptualization process of the computer-aided feedback (i.e., operationalization from manual to automated data processing in Jira) [126, 134, 137].

The fact that the data capture concept builds on technical methods that extend an existing database management system, such as in Jira, reduces the possible loss of quality within the data. The type of data acquisition also plays a significant role (e.g., the applied digital surveys do not allow outliers or noisy data). While manual questionnaires have previously led to incomplete data or ambiguous ratings, the digital survey is ideal for consistent data quality (e.g., a yet incompletely answered survey does not allow to be submitted) [207].

Data integration and reduction are applied to merge, reduce, and integrate data from different sources into one central database with categorical representations through a table (e.g., the time-related cluster within the socio-technical data based on the different capture methods and times). The reduction of data attributes and tables reduces the degree of heterogeneity with relevance to the data management system. Thus, the feedback processing (e.g., moods of single team members into aggregated team values without violating the attributes meaning) decreases the number of data features, which is strongly advised when working with statistical analyses and machine learning models [38, 221]. The preprocessing routine automatically merges and reduces new socio-technical data every week. For example, after the preprocessing, a table called "weekly analytics" stores the socio-technical data at a representative team level, allowing consecutive feedback processing (e.g., dependency analyses, modeling, and visualizations) to access prepared data quickly.

Figure 5.2 shows an example of the weekly data preprocessing. The raw data pre-processing with significantly more entries are essential for subsequent analytical operations.

Mood-Related Data Table (*Source Survey*)

id	project id	sprint id	member id	time epoch	panas active	panas nervous	panas alert	...
1	11000	168	199280	1618237463	3	3	2	...
2	11000	168	199281	1618237473	2	2	3	...
3	11000	168	200313	1618237483	2	3	1	...
4	11000	168	223833	1618237493	3	2	1	...
5	11000	168	233832	1618237503	2	3	2	...
...	...	...	...	...	...	...	...	...

Weekly captured mood data by team member

Task-Related Data Table (*Source Jira*)

id	project id	sprint id	member id	time epoch	scheduled tasks	competed tasks	cancelled tasks	...
1	11000	168	199280	1618237463	4	1	3	...
2	11000	168	199281	1618237473	10	6	4	...
3	11000	168	200313	1618237483	7	3	4	...
4	11000	168	223833	1618237493	4	4	0	...
5	11000	168	233832	1618237503	7	3	4	...
6	11000	168	199280	1618323863	8	5	3	...
7	11000	168	199281	1618323873	10	9	1	...
...	...	...	...	...	...	...	...	...

Daily captured task-related data by team member

Weekly Preprocessed Socio-Technical Data Table

id	project id	sprint id	calendar week	panas positive	panas negative	scheduled tasks	completed tasks	cancelled tasks	...
1	11000	168	15	3	3	50	27	23	...
2	11000	168	16	2	2	43	33	10	...
3	11000	169	17	2	3	38	32	6	...
4	11000	169	18	3	2	41	38	3	...
5	11000	169	19	2	3	13	10	3	...
...	...	...	...	...	...	...	...	...	...

Weekly clustered task and mood data of one teams

Figure 5.2: Preprocessing: Data Merging, Clustering, and Transformation

Data transformation and discretization discretization are of particular relevance in this concept because some statistical operations and machine learning methods require the socio-technical data to match certain constraints before processing (e.g., normalized value ranges of data attributes). However, for distinctness or time-based ranking operations, the raw data Unix timestamps in seconds are sufficient, while weekly clustering of multiple days works better using calendar weeks.

Simple data attribute transformations, as shown in Figure 5.2, are applied to better organize the data and improve its accessibility for both humans and machine routines (e.g., transforming a Unix timestamp "timeepoch," which represents the seconds that have elapsed since January 1st in 1970, into minutes, hours, days, or even into a calendar week). All transformations in this work preserve a particular data attribute's meaning. The primary data transformation in the preprocessing routines facilitates its compatibility for consecutive feedback processing based on preformatted and validated data accessible from a central database (e.g., derive descriptive, predictive, and explorative sprint feedback based on the weekly preprocessed data foundation).

The preprocessing ensures that the database stores only the most relevant data attribute formats with global relevance for the different feedback types. The individual feedback processing routines require specific data transformations to fulfill the application purpose (e.g., input formats for machine learning models and normalized data attributes for exploratory analyses). Descriptive feedback is primarily based on aggregations and statistical operations (e.g., identifying anomalies [169]). The predictive feedback accesses the same weekly preprocessed data foundation but requires different transformations to fit the formats for building and validating machine learning models [38]. The explorative feedback uses advanced statistical operations that require yet another data format [134, 194].

Consequently, the specific feedback processing described in the following section determines the exact data formatting needed to provide the different computer-aided sprint feedback types based on the centrally preprocessed socio-technical data foundation. Computer-aided sprint feedback addresses this thesis's overall goals and is conceptualized according to the following three values:

⇒ *Simplicity:* usage and maintaining

⇒ *Supplementation:* reliable, supportive, knowledgeable

⇒ *Integration:* consistent with agile methods and practices

The preprocessing step ensures the data quality. It protects the feedback methods and models from potential null values, duplicated entries, false indexing, and even incompatible formats. It is essential that this preprocessing has been conceptualized as add-on support, meaning that original database structures and interfaces remain unchanged.

5.2   Descriptive Sprint Feedback Asset

The retrospective feedback processing routine supplements the conversion of preprocessed socio-technical data into a more application-specific form. It makes the preprocess data foundation more meaningful and supports information in projects about team dynamics in sprints through additional descriptive, predictive, and explorative feedback. Moreover, the retrospective feedback aims to support the understanding of socio-technical aspects for future sprint planning and decisions derived from teams' perceptions and project data, reflecting the ideology of the agile method Scrum [209, 220].

In Scrum, the software release goals are defined in backlogs, portioned into manageable development goals during the sprint planning. The fast delivery of software fragments through short development iteration makes progress highly manageable. It is thus easily recognizable whether the work is oriented correctly. In particular, it allows the team to review evident achievements or failures, consider different methods and teamwork aspects, and identify areas of improvement. With every completed sprint, the team shares their experience to inspect and adapt their teamwork and development methods in the sprint retrospectives [59, 180].

The retrospectives enable whole-team learning, act as a catalyst for change, and generate actions. Retrospectives go beyond perfunctory sprint closures and focus on the development process to assess the scheduled sprint goals' fulfillment corresponding to the previous weeks' objectively tracked development performances and team member reflections. Therefore, sprint (progress) reports are commonly used as the retrospective data foundation for the last sprint [63, 186]. Nevertheless, it is also a team activity that identifies what worked well and what needs improvement based on recognized problems or suboptimal situations (see Chapter 2.1.2).

Software development is an activity that involves humans [65, 198]. Therefore, sociological aspects (e.g., interpersonal tensions) are essential to identify and understand as technical issues [59, 238]. Social factors are often insufficiently addressed in software projects, including the systematic capture of human factors and the characterization of relationships between team behavior and sprint outcomes. Proper information support during a retrospective meeting is crucial to a holistic understanding of previous performances and produces helpful knowledge for the next sprint planning activity [59].

A international survey study about the information need in ASD was conducted with practitioners and researchers with domain expertise [138]. Among several other questions, the participants were asked about their experiences on teams' frequent information usages and problems in sprints. For example, the three most frequently stated reasons for lacking sociological information usages in teams during sprint planning were as follows:

> *"not available or no use."*

> *"extra time, cost and effort."*

> *"lack of tools or knowledge about it."*

These reported reasons are unnecessary and avoidable considering technological feasibility, like this thesis' computer-aided sprint feedback concept. However, as previously described in Chapter 4, the currently leading management systems on the market, including Jira, primarily concern tracking and optimizing task-based development progress and subsequently derived measurements such as the velocity of teams. Sociological effects, such as overly ambitious sprint goals that might result in team exhaustion, are rarely systematically captured or investigated.

The computer-aided sprint feedback described in this chapter is conceptualized to address the above-reported information gaps and provide a fast feedback processing that practically supports understanding team dynamics in ASD. The retrospective sprint feedback covers weekly updated visualizations of socio-technical aspects in sprints and descriptive analyses for determining trends and anomalies in the team behavior (e.g., a substantial deviation from the current communication behaviors compared with previous weeks). The concept is complementarily established as the prototype for Jira.

The data analytical processing concept involves accessing the preprocessed socio-technical data (e.g., weekly intervals using a scheduler routine or manually started by a Jira admin) without further manual operations to descriptively analyze the previous week's team behavior (e.g., detecting abnormal deviations). Figure 5.3 depicts the retrospective sprint feedback processing using a FLOW notation to characterize the included roles, resources, tools, and information outcomes [207].

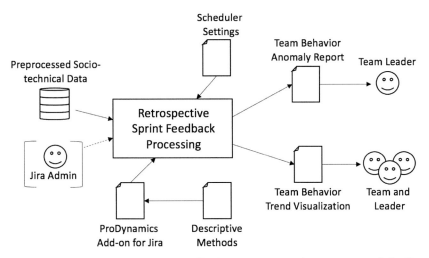

Figure 5.3: Retrospective Sprint Feedback Processing as Flow-Activity , cf. [207]

The team behavior feedback covers four aspects of team dynamics in sprints –communication, mood, satisfaction, and performance – based on the GQM understanding goals previously defined in Chapter 4.3. The retrospective feedback processing results in two outcomes: a team behavior anomaly report for the team leader and team behavior trend visualizations accessible by all team roles. The anomaly report is conceptualized as a systematic warning mechanism that sends an email notification to the leading team member about unusual behavioral occurrences during the last week [169].

5.2.1 Team Behavior Courses and Anomalies

Retrospective feedback about courses and anomalies of team behavior and the development performance (e.g., based on socio-technical data) supplied valuable insights and promoted insights applicable in future sprint planning. The descriptive methods in Chapter 2, the mean (see Equation 2.5), median (see Equation 2.6), standard deviation (see Equation 2.8) and linear regression (see Equation 2.9) present simple statistical methods that allow characterizing team behavior changes over time and detect abnormal occurrences (e.g., unusual activities). The retrospective feedback uses automated anomaly detection through descriptive methods [169]. The manual derivation of regressions is a tedious process, and systematic processing routines handle them more efficiently. Figure 5.4 shows the descriptive methods applied in a practical example of the varying number of team meetings held during a week in a software project.

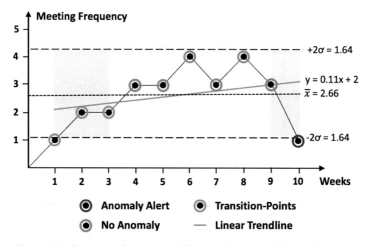

Figure 5.4: Course and Anomaly Characterization: Meeting Behavior

Course: Involving a linear regression line for time-series data supports depicting the overall course of change (e.g., the meeting frequency trend based on the last ten weeks) [106]. Trend lines are not always suitable for predictions, as shown in Figure 5.4. Nevertheless, trend lines are helpful as a directional indicator in projects, characterizing positive, negative, or constant behavioral courses (e.g., overall decreasing team mood). In the example, the regression equation used to describe the number of meetings y indicated a slow increase over time. Since the first week, the meeting course positively changed by a factor of .11 and a general intercept of +2. In ASD, teams usually require time to settle in, resulting in fast behavioral changes and improvements at the beginning of a new project [62]. Thus, with every consecutive week, courses have to be updated. In this retrospective feedback concept, the provided trend lines cover a limited time range of maximal ten previous weeks or sprints to avoid including outdated behavioral data. Restrictions have also do apply for detecting anomalies.

Anomaly Detection: Concerning the visualization of sprint feedback based on time-series data, anomalies and outliers are sometimes cognitively identifiable in data observations when values significantly differ from the rest of the data [126, 135, 221]. However, anomalies are not always clearly recognizable or are subject to different interpretations. Moreover, there is not always enough time to look for anomalies during everyday development routines. Therefore, the retrospective feedback concept in this work also considers the computer-aided recognition and automatic notification of behavioral anomalies based on descriptive statistics. The use of more complex methods (e.g., detecting anomalies through machine learning) was not used, as each team's behavior has to be individually investigated. Thus, the information source based on a team's weekly data is too small for reliable findings [169].

For automatic anomaly detection through statistical methods, it is essential to determine irregularities as accurately as possible (i.e., avoiding false positives). Mircea's [169] anomaly detection method was adaptively used in his concept to reduce false-positive anomaly alerts with the following case distinctions:

1. Settle-in period of three weeks for new teams/projects

2. No alerts for back-to-back anomalies

3. Anomaly analysis concerning the last ten weeks

Figure 5.4 depicts the first three weeks in the meeting frequency example, which are not considered transition points within the anomaly detection routine. The follow-up weeks until the ninth week are descriptively analyzed to obtain the standard deviation σ ((e.g., up to the maximal time sequence of the previous ten weeks). On a five-stage interval scale, only the meeting frequency values near the mean $\bar{x}$ are not identified as anomalies (e.g., the fourth, fifth, seventh, and ninth week). Consequently, the anomaly report results in two false-positive alerts for the sixth and eighth week, with only one true-positive alert for the tenth week.

The concept includes a variable dependent range so that the accuracy of true-positive anomaly detection increases (e.g., with doubled standard deviation) [169]. In contrast, significant anomalies are found before more minor irregularities between the $\pm\sigma$ and $\pm 2\sigma$ range [169]. Another optimization is the tolerance for occurring back-to-back anomalies (e.g., meeting frequency value one in the eleventh week). Suppose a meeting frequency value suddenly decreases or increases significantly (e.g., due to the beginning of the holiday season). In that case, the dispersion measure requires at least one follow-up week to adjust to the new level, and another alert is avoided if there are two consecutive outliers [169]. The weekly anomaly detection routine automatically processes each socio-technical metric defined in the GQM-model (see Appendix C.4).

Weekly anomaly findings are summarized in a report and automatically send by email to team leaders using the same email notification service as realized for external surveys in Chapter 4.4.2. The anomaly notification provides for each finding a descriptive implication and link to the referring time-series data visualization in Jira. A notification example is shown in Appendix D4.1. The retrospective visualization concept for the socio-technical sprint data is described in the following section.

5.2.2   Visualization of Socio-Technical Sprint Data

The retrospective visualization is conceptualized through five socio-technical sprint feedback modules. Each concerned the previously defined GQM goals for understanding team dynamics in sprints based on the captured team communication, mood and satisfaction, performance, and task-related interactions. For reasons of compliance, the graphic design and color scheme followed a style guide for Jira software development[1] that was also applied to the visualization concept of the predictive and explorative sprint feedback support (see Section 5.3 and 5.4).

The modules strive for feedback simplicity, supplementation, and integration into the professional working environment agile teams using Jira. They are assessed and refined through an overall evaluation of the industry concerning the usability and utility of the computer-aided feedback described in Chapter 6. The maturity of the feedback visualization concepts emerged iteratively during the different research stages shown in Figure 5.5.

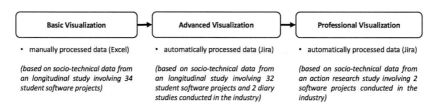

Figure 5.5: Iterative Visualization Stages of the Feedback Modules

The initial sprint feedback concept involved basic visualization techniques to understand team behavior based on manually captured data in 34 student software projects (e.g., line charts and MS-Excel communication networks) [35, 126, 135, 207]. A subsequent, more advanced feedback visualization emerged from a technology transformation (e.g., considering web application visualizations) [36]. The advanced feedback techniques supported the visualization of socio-technical aspects in ssprints, focusing on student software projects [134, 136, 139]. The professional visualization presents the latest refinement of the advanced feedback concept through practitioner feedback concerning the usability and utility of the computer-aided sprint feedback and style guides of Jira (see Chapter 6).

---

[1]https://atlassian.design/

5.2.2.1   Sprint Board Activity Module

The first retrospective module in this sprint feedback concept supports reviewing task-related activities over time (e.g., daily progress of single issues processed by the team members). The module is developed to provide a reviewable sprint board that animates activities by the time [139]. In contrast, conventional sprint boards depict the current sprint progress but do not depict the progress changes and actions made in the past [155]. The system logs all team-related task activities so that the retrospective player primarily loads logged states of every sprint task. The retrospective player module processes the transition states of past tasks involving timestamps and associated background information (e.g., who worked on it from when and issue types).

Retrospective module maps the process data onto the retrospective player (i.e., sprint board with the time axis). Therefore, the sprint board player is conceptualized to dynamically adapt the visualization corresponding to teams' development workflows (e.g., an extended process: To-Do $\rightarrow$ In Progress $\rightarrow$ In Review $\rightarrow$ In Test $\rightarrow$ Done). Figure 5.6 shows an example of the provided sprint board activity visualization with a standard development workflow in Jira.

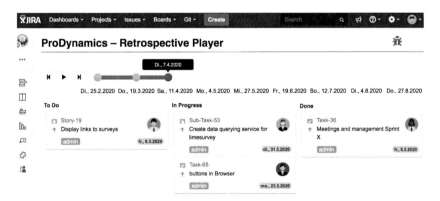

Figure 5.6: Retrospective Module: Sprint Board Player

The retrospective module supports the systematic review of team conflicts and task-related problems (see Section 4.3) by tracking textually annotated issues of a specific task during the sprint retrospective or directly in an active sprint board [139]. The module provides a shared asset for reporting task-related issues, accessible by all team members at any time. Moreover, it systematically supports the quantitative processing for further investigations, such as analyzing whether an increased number of reported task-related problems correlates to a lower development velocity or intensified opposing team mood in a specific sprint. The latter is essential for automated anomaly detection and predictive feedback through machine learning models and the explorative feedback on socio-technical sprint dependencies (see Section 5.3 and Section 5.4).

5.2.2.2   Team Communication Module

The second retrospective feedback module for team communications in sprints satisfies the quality focus and goals (*G*2) to support shared understanding of (dys)functional structures and communication behavior changes (see Appendix C2.1). Varying team members' perceptions about the overall communication situation within the team are made accessible as aggregated team values by the weekly retrospective processing routine through the visual representation of the systematically captured reflections of individuals. Moreover, the visualization of team communication characteristics on a timeline, as shown in Figure 5.7, forms a shared basis for team discussion in sprint retrospectives.

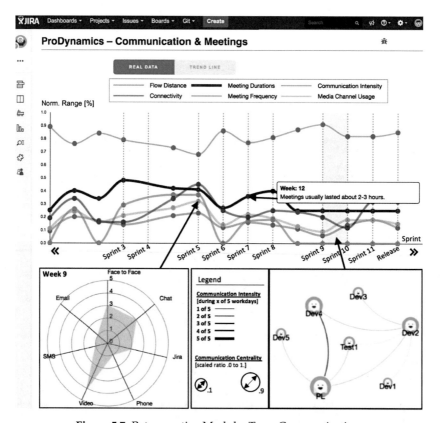

Figure 5.7: Retrospective Module: Team Communication

Understanding teams' communication behavior (e.g., use of media channels and structures) helps identify the regular course of the team and irregularities. Timeline visualizations provide teams' comparable communication behaviors retrospectively (i.e., in this concept, up to ten sprints of weekly data). The interactive concept allows a custom selection of specific sprint periods to be visualized.

The information technologies used for computer-aided sprint feedback come with particular advantages. One such benefit is that all modules preserve encrypted intra-team data, and for single members' concerns, an anonymity mode automatically veils the identity of all team members' information. Another relevant benefit is the support of interactive information features covered in the team communication visualization concept. The following visualization concept fulfills the defined goals $G2$ for understanding team communication in sprints (see Appendix C2.1).

Visualization of Communication Networks and Courses

The line chart visualization is applied to summarize and compare team communication courses in a timely order for a maximum of ten previous sprints. For the y-axis, a normalized value range (0 to 1) was used for the six communication metrics listed in the legend (see Chapter 4)), identifying changes in the course of the individual metrics more easily. A toggle button supports the view change regression-based trend lines of the observable team communication (see Chapter 2). The latter determines the tendencies (e.g., whether the meeting duration was stable during the last ten sprints or whether there was a downward trend in communication intensity). Simple forward-backward sprint navigation enables the display of different sprint sequences (e.g., to review a specific situation from the past).

The timeline visualization complements the systematic anomaly detection described in Section 5.2.1 and supports the statistical finding of irregularities beyond the usual team communication behavior. However, the advantages of using information technologies include the accessibility of additional details. The line chart covers hover and select interactions providing extensive information about a metric.

- **Hover function:** This function provides different details depending on the hovered metrics definition. For example, when hovering over the "meeting duration" line, extra information provides a textual description of the team's usual meeting duration over the course of 12 weeks. In contrast, hovering over the "media channel usage" line provides more complex information. A radial chart depicts the team's communication media usages in a particular week on the five-point intensity scale, shown in Figure 5.7. The additional detail reveals that "video"-communication was more intensively used than other channels.

- **Sprint selection function:** This function provides several details about the communication structures in teams and sprints based on social network analyses. In the example shown in Figure 5.7, the selection of sprint 10 leads to a complementary team communication network characterizing the centrality of the team member, interconnecting communication intensities, and identifying loners. The graph properties are automatically scaled (e.g., indicating the communication diversity within the team). Therefore, the users (by name or anonymized by roles) cover additional textual values undermining statuses (e.g., maverick characteristics) based on a comparable communication behavior with others.

5.2.2.3   Mood and Satisfaction Module

A third retrospective feedback module satisfies the previously defined quality focus and goals (see G1 and G3 in Chapter 4.3) to support the shared understanding of changing moods in teams and the awareness of external team performance perceptions in sprints. The feedback is visualized using a line chart depicting the teams' previous courses of positive and negative mood effects (PANAS-SF) on a timeline [225]. An overview of the perceived team performance based on external perspectives complements the latest sprint feedback for teams in this module. This module considers both customer and team leader reflections.

Visualization of Positive and Negative Mood Courses

The main visualization of this module is again realized using a timeline chart, characterizing teams' commonly positive and negative mood levels and deviations during the development weeks (in sequential order for a maximum of ten sprints). At first glance, the visualizing concept shown in Figure 5.8 seems similar to the previously described team communication module. Nevertheless, it holds some remarkable differences – for example, all displayed mood values present aggregated team measures.

The positive and negative mood values are derived using the median, considering each team members' preprocessed mood reflections from weekly behavioral survey data (see Chapter 2.1.5). The colored offset areas (e.g., the green-colored increase at the eleventh week) visually highlight deviations up to ten displayed sprints compared to the median mood level. Deviations do not imply anomalies, which are complementarily analyzed by the systematic anomaly detection described in Section 5.2.1. The y-axis covers the perceived mood intensity range from one to five, originating from the positive and negative mood affects scale (see Chapter 4). Additional details are provided through hover and select interactions within the line chart. In Figure 5.8, hovering over the negative mood line in the sixteenth week, a radial chart depicting the different moods affect the characteristics of the adapted PANAS-SF items corresponding to the teams' perceived intensity [207, 225]. The radial chart allows a differentiated interpretation of the single mood properties, in this case, an increased fear in the team before relevant events.

External Team Performance Satisfaction

The specific sprint selection in Figure 5.8 leads to additional feedback about the team performance perception from external perspectives, provided within the shown select menu on the right. The first two menu sections cover the captured external performance ratings at the end of sprints (see Chapter 4.4.2)—the responses concerning perceived team performance measures about the software and the team. A comparison of systematically captured perspectives from different roles (e.g., the team leader is the standard selection) determines the satisfaction of expectations based on the accomplished sprint performance (i.e., complementary to the objectively measured fulfillment of development task). The third menu section covers some background information of the reflecting person (e.g., self-rated experience in software projects and as a customer), based on a similar survey study conducted by Schneider et al. [207]. The visualization of team moods on a timeline promotes a shared basis for team discussion in the sprint retrospectives, similar to the previous team communication module (e.g., sprint performance outcomes compared with changes in team moods).

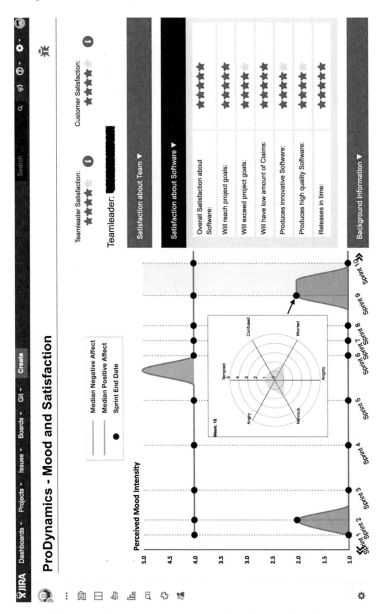

Figure 5.8: Retrospective Module: Mood and Satisfaction

5.2.2.4 Productivity Comparison Module

The fourth module in this retrospective feedback concept satisfies the shared understanding in teams of development performance variances ($G4$) within and across sprints. With the native support of agile software projects, current management systems provide tracking and visualizing development progress measures in sprints (e.g., generated burndown charts, velocity diagrams, and other reports related to task states) [63, 153, 155]. In contrast, but not necessarily in a revolutionizing manner, the described module was conceptualized to complement retrospective feedback concerning the specific understanding goal (G4), the related quality focus, and comparable performance measurements (i.e., progress and workload), as described in Chapter 4.3. This module's visualization concept follows a similar information representation scheme to the previous two modules and compares the productivity changes.

Visualization of Performance Variances in Sprints

The main visualization of the productivity comparisons module builds on a timeline chart depicting a team's objectively measured development performance for a maximum of ten completed sprints (i.e., based on the task-related progress tracking, in this work, by Jira). The objectively measured development data is processed into comparable and interpretable performance feedback provided through visualizations of aggregated task measures, as shown in Figure 5.9 (e.g., course of velocity changes and task-load balance in teams). The line chart depicts the development performance changes in sprints, allowing the direct comparison of task-related productivity measures. Therefore, the line chart has a multi-axis scaling to map different value ranges of the task-related attributes on the same time axis (e.g., done issues together with logged hours). Figure 5.9 shows a performance observation based on one of the industrial software projects described in Chapter 6.

The line chart reveals a strong alternating development performance outcome during the first four sprints, recognizable by the alternating number of "done story points" (light-green) or the velocity course. The velocity resulted in the percentage of the total number of "done story points" (i.e., including pruned, modified, and added story points in an active sprint) compared to the original "estimated story points" during the sprint planning. However, alternating development velocities in the first sprints are not unusual and often observable in newly formed teams or at the beginning of a new project [63]. A fast and steady velocity within a short period should be the target because it indicates the reliability of the sprint estimation based on a teams' possible development performance.

Diverging Stacked Bar Chart Visualization for Performance Variances in Teams

Further details about performance divergence are provided through hover and select functions (e.g., textually and visually depicting the performance characteristics in a particular sprint). For example, when selecting the "Sprint 6" shown in Figure 5.9, the diverging stacked bar chart characterizes the varying development performance and workload balance within the team for a selected sprint [36]. Depending on the team member's privacy preferences, the performance deviations are visualized anonymously by project roles. The timeline visualization and the additional diverging workloads and development performance within teams allow a more granular understanding of the accomplished velocity in sprints. Thus, it provides a discussion foundation in sprint retrospectives concerning the ratio between logged hours for completing tasks and the total number of completed tasks. However, it is essential to mention that all the covered task-related data in this module are natively accessible by team members in Jira and processed into an aggregated, more time-comparable form.

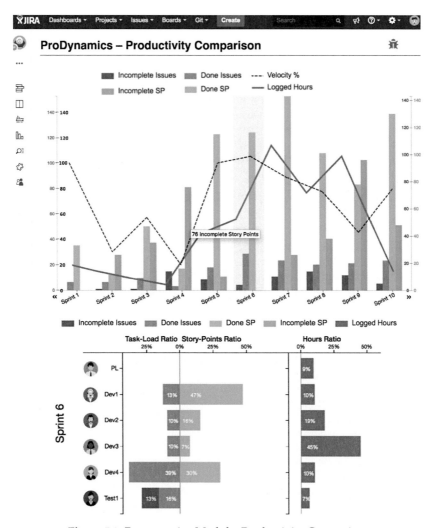

Figure 5.9: Retrospective Module: Productivity Comparison

5.2.2.5 Interactions Revealer Module

The fifth retrospective module concerns the feedback support for a shared under-standing in teams for interaction networks based on task-related measurements. In contrast, the previously described retrospective feedback module for team commu-nication characterizes behavioral and structural changes in sprints solely through team member reflections (i.e., the subjectively rated communication with others). Consequently, the responses make the team communication module highly depen-dent on team members' individual motivation to participate (see Chapter 6). In contrast, this retrospective feedback module on the team interaction networks in sprints considers only objective data from task-related activities commonly tracked within agile project management systems. Nevertheless, the exact task-related in-formation available for feedback processing can vary, corresponding to the used project management systems.

For example, natively tracked activity information associated with develop-ment tasks in Jira include attachable files (downloads/uploads), comments (writes/reads), task (re)assignments (forwards, completions), and status notifica-tion (personal or others) [149]. These four activity types are systematically tracked for each development task, a natively available data source in sprints. At the end of a sprint, the module quantitatively analyses the activity data for associations be-tween team members – for example, "User A" attached a file to "Task 12" in "Sprint 1", which was accessed in the same sprint two times by "User B," one time by "User C," and never by "User D".

- **Assignments:** The activity is triggered when a task has been created or (re)assigned to a team member. Self-assignments are not considered since they do not involve interactions with other users. The number of assignments depicts how involved the developer was during a sprint compared to other team members (e.g., the amount of assigned work or forwarded tasks).

- **Comments:** Every development task supports adding personal or shared tex-tual comments. A task-related comment contributes to the information shar-ing by teams (e.g., advice, problems, customer feedback, and issue details). In Jira, teams use the directed command "@user" to notify a specific member.

- **Watcher:** Notifications are sent to an assigned team member in case of changes. By default, an active watcher-event triggers when a task is modi-fied (e.g., including status changes corresponding to the development work-flow, new comments, attached files, and (re)assignments). Tracking the issue ensures that the activities linked to the task are observed.

- **Downloads:** Task attachments are primarily relevant for the assigned mem-ber but also accessible by others. This includes sharing solid information (e.g., attached documents, media files, and code snippets) [149, 207]. Track-ing attachment activities identifies whether provided information by a person found proper consideration by others.

In this concept, the four activities quantitatively analyze all task-related interactions, allowing for visualizing team interaction structures based on the social networks described in Chapter 2. Moreover, the visualization concept from the team communication module is adapted to visually characterize the task-related team interaction in sprints in a comparable form. The concept depicts interaction structures that emerged from the tracked task-related activities connecting two members. Task-related activity tracking requires no external measurement methods (e.g., opinion polls). However, the concept builds upon a teams' active Jira usage. Figure 5.10 shows an example of a derived team interaction network in the prototypical extension for Jira systems.

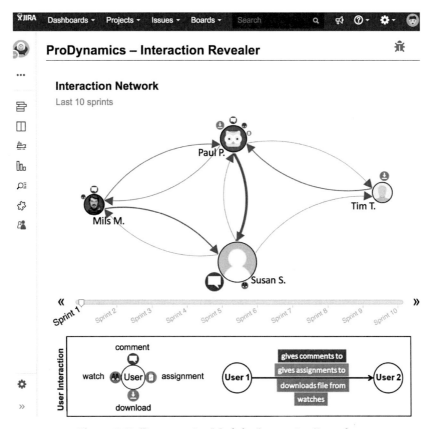

Figure 5.10: Retrospective Module: Interaction Revealer

The scaled connection-width between two team member nodes results from the total analyzed (directed) interactions (i.e., considering the four tracked task-related activities). A scaled node size depicts each team members' relative centrality based on the ratio of all weighted in- and outgoing interactions compared to other members (sub-nodes respective to the total interactions of an activity type). The interaction network underlines the possible understanding support through processed and visualized feedback based on natively available data.

5.3   Predictive Sprint Feedback Asset

Current project management systems, that support agile teams, usually provide simple prediction methods or trend lines concerning the estimated development progress for the remaining days in sprints, as in burndown charts [155]. Although these future trends are systematically derived, they are often limited to conventional statistical analyses, such as linear regression or mean values). Moreover, these analyses are not always the most suitable methods for considering changing development behaviors in teams [137]. As a part of this work, a conducted survey study has shown that agile teams frequently lack the time, resources, or knowledge for applying analyzing methods investigating human factors in ASD [138].

The retrospective modules presented in Section 5.2 of the computer-aided sprint feedback concept provide the foremost descriptive information processing and visualization support for teams on the behavioral changes (i.e., based on socio-technical data) over time in agile software projects [139]. Consequently, team behavior patterns and development courses are made recognizable and disclosed through past trends. However, sophisticated behavioral patterns and predominantly interdependent socio-technical aspects are not always noticeable through timeline visualizations, which can affect the interpretability of concrete tendencies for the next project weeks. Therefore, a prediction module is developed to support the understanding of future team behaviors based on the covered socio-technical aspects (see Section 4.3).

Machine learning (ML) disclosed a growing interest in recent years for prediction and decision support systems in various application domains, such as software engineering and clinical medicine [27, 86, 232]. The increasing demand has also been noticeable in non-professional fields requiring ML systems that are ready-to-use or easy-to-operate, such as automated machine learning techniques [73, 105]. However, developing ML applications is challenging for others than data analysts because the systems must be operable without expert knowledge and quickly add value without significant efforts. The utility of the predictive module for computer-aided feedback about the team dynamics in sprints depends on the practical simplicity, information supplementation, and integration of the prediction support.

The predictive module in this computer-aided sprint feedback concept covers an automated data processing routine. The processing is crucial for the prediction performance of ML models based on qualified and transformed data and automated feature selection (AFS) for optimal data attributes when training models. For a maximal prediction performance of the following weeks' team dynamics, an automatic selection of the most suitable ML model type is realized. The selection option cover a set of five supervised learning algorithms covering different learning techniques applied to past socio-technical sprint data [73, 105]. Each algorithm was included according to its relevance for the predictive sprint feedback module. Figure 5.11 shows the processing routine of the ML prediction model. The socio-technical data foundation is made amenable to the ML-model ensemble using a shared data format and AFS method for the best prediction precondition [38, 226].

The five ML algorithms are based on supervised learning that analyzes the ML input data and generates an inference function, which is used to map new weeks' team behavior. In supervised learning, each instance is composed of an input object and an expected output value (see Chapter 2.2.4).

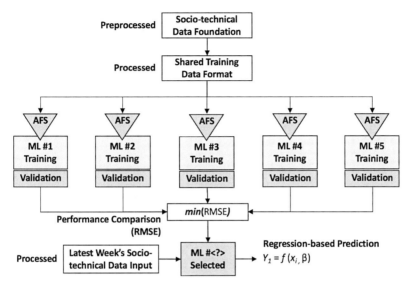

Figure 5.11: ML Prediction Processing with Automated Feature Selection

Based on the ensemble of included ML algorithms, different regression techniques are used to analyze the relationship between a dependent variable and the independent variables, as well as the degree of influence of multiple independent or explanatory variables on a dependent variable [221]. While classification problems are used to label an instance, usually, the result is a discrete value [226]. Regression problems need to predict a quantity, and the input variables of the regression can be continuous or discrete [38]. Consequently, this predictive module is included to analytically derive regression-based future trends for a target attribute (dependent variable) represented through a functional expression that uses the explanatory variables as their function inputs.

The final cross-validation of each ML model's prediction accuracy determines the overall performance. Thus, the best model is selected for the next week's prediction. The weekly predictions are available for teams through an interactive line chart visualization in Jira. The five ML algorithms in the concept were implemented in Jira systems using the open-source machine learning library, *Waikato Environment Knowledge Analysis (WEKA)*. The following subsections describe the shared training data format, included ML algorithms, AFS, applied performance validation, and the visualization concept of the predictive module.

5.3.1   Regression-Based Machine Learning Algorithms

The predictive module's concept considers five regression-based ML algorithms applied to an automated model selection routine that maximizes the weekly prediction accuracy. It does so by finding the most suitable regression model for interpreting the dynamically changing socio-technical dependencies in the data. The regression models aim at learning the relationship between a feature set of explanatory variables and the target variable to be regressed. The covered regression models rely on unique techniques, and each comes with strengths and weaknesses regarding the interpretation of relationship properties in the applied socio-technical data set. Different regression-based problems cannot be solved consistently by a single, static prediction model. The following five ML algorithms were included in this predictive module based on their technical capabilities in functionally interpreting the different socio-technical relationship properties in the provided data foundation (see Chapter 4).

The focus of this module is the practical application and integration of existing ML algorithms provided in the open-source WEKA library. Several regression analyses were covered based on different ML techniques, supporting a higher interpretation capability of socio-technical aspects at comparable prediction accuracy [238]. The description of the included ML algorithms was kept simple because the functional details of each would have exceeded the ML introduction scope in this thesis. Moreover, all five algorithms are well documented in the existing ML literature [9, 38, 66, 221, 240]. In addition, functional details of the included ML algorithm with practical examples can easily be found online [2].

Multiple Linear Regression (MLR)

MLR is included in this module to support the characterization of linear dependencies among the socio-technical aspects captured in sprints, thereby deriving linear trends for the underlying measures. The MLR uses multiple variables to predict a dependent variable, also known as regression problems [221]. An MLR that only considers one independent variable is a simple LR (see Section 2.2.4). However, an MLR model performs best for predicting significant linear relationships between the independent variables and the dependent variable (e.g., scheduling the next sprint's development tasks according to the number of available developers and working hours). An MLR model estimates linear trends of a dependent variable regressed through multiple independent variables [38]. MLR results in a simple regression model that supports fast training and an accurate prediction performance if the output variable is based on significant linear relationships of the inputs [221]. The model is easy to update because it requires no parameter adjustments and feature scaling. Moreover, the derived regression function (model) is comprehensive (e.g., possible to derive manually). However, the MLR's solely linearity analyses are not suitable for non-linear relationships because the functional properties are only insufficiently interpreted [194].

---

[2]https://www.tutorialspoint.com/weka

K-Nearest Neighbors (KNN)

KNN is included in this module because it enables predictions on socio-technical metrics independently to linear or non-linear dependencies within the applied training data. KNN is a non-generalized machine learning algorithm without assumptions on data nor performing statistical learning. Thus, new data is added without impacting the algorithm's accuracy [168]. Instead of involving statistical analyses, KNN's perform data value comparisons for finding the K most similar instances within the training data set that have the closest distance to the new input instance, thereby predicting the target variable [9]. Different Ks might lead to changed regression results. Consequently, this module covered an automated K-selection that iteratively finds the K accomplishing the overall highest prediction accuracy for the training data [168, 226]. Due to the different value ranges of the socio-technical data features, a min-max normalization is performed so that the feature variables' values are considered equally weighted in the KNN search [148]. The KNN algorithm is included to predict numerically (unlabeled) socio-technical metrics [135, 221]. The regression-based predictions is derived for the new input data based on the mean value of the K-nearest instances from the training data. Therefore, the KNN measures the Manhattan distance between different feature values [31]. The Manhattan distance is applied because of the different socio-technical data types. The returned mean value of K-nearest instances is the corresponding prediction result. A significant disadvantage of KNN is its inability to provide the data's inner meaning and hidden rules.

Decision Tree (DT)

DT is included because it is a non-parametric prediction model that uses binary rules to represent the relationship between attributes and the target variable. Neither the previous LR nor KNN techniques derive optimal predictions for non-linear data relationships. DTs refer to Classification And Regression Tree (CART) and are efficient at acquiring non-linear relationships and mastering the feature interaction in data sets [227]. The DT was applied as a regression tree for non-linear separable socio-technical data and analyzed the probability for prediction (continuous) outcomes. DT models handle numerical and categorical variables without feature scaling requirements and are composed of nodes (socio-technical feature), leaf nodes (values), and directed edges [9, 168]. In this module, the DT was applied for regression tasks of the socio-technical training data features, which is tested starting from the root node. The DT recursively tests and distributes the sample until reaching the leaf node to make predictions (i.e., a new instance will be assigned to the next level nodes, where each node corresponds to a value of the following feature) [57, 157]. Consequently, the DT algorithm in this work covers an automated depth finder of the regression tree and pruning to avoid overfitting, thereby improving a model's generalization for new instances [226]. An advantage of regression trees is the interpretability of predictions (e.g., a visualized tree depicting decision paths). However, regression tree models are not easy to update when adding new data and may result in overfitting. Moreover, a small amount of changing data can lead to an entirely restructured DT.

Support Vector Regression (SVR)

SVR is included because it supplements the predictive module's interpretation capability of more complex data relationships, which otherwise would not be suitably identified by the previously described three ML algorithms (e.g., functional relationships in the data). SVR models are a specific adaptation of the Support Vector Machine (SVM) [215]. SVR finds a regression hyperplane of instances in the training data that are the closest to the plane to minimize the cost function error [64]. In contrast, SVM is a supervised binary machine learning model that defines a hyperplane (decision boundary) with the most considerable interval in the feature space that best separates training data into different categories. A key parameter in SVM and SVR is the kernel function. The kernel function performs operations on the low-dimensional space before converts the features from low-dimensional to high-dimensional. Therefore, SVM and SVR algorithms can use different kernel functions, such as linear, non-linear, polynomial, radial basis function (RBF), and the sigmoid function [12, 215]. The RBF kernel is used for the SVR to transform the socio-technical training data to a higher dimension (vector space), approximating the function of complex data dependency shapes. The selection of kernel functions makes SVR flexible for solving various non-linear regression problems. An advantage is that it solves the regression problem with a relatively high prediction accuracy and strong generalization ability [215]. In contrast, it is susceptible to missing data. Moreover, there is no universal standard for choosing a suitable kernel function when investigating non-linear problems.

Multilayer Perceptron (MLP)

MLP complements the ML model-based prediction support of the four previous algorithms. In this work, MLP is used to solve non-linear regression problems. MLP models are a class of artificial neural networks with a feedforward structure (i.e., mapping input vectors to a set of output vectors) [80, 163]. An MLP is viewable as a directed graph that consists of multiple node layers, where nodes are fully connected to the next layer [80, 181]. Except for the input node, each node is a neuron with a learned non-linear transformation function that maps input samples to a dimension where the input values are linearly separable. The non-linear transformation function is called the activation function. The most commonly used are the following: Sigmoid function, Tanh function, rectified linear unit (ReLU), Leaky ReLU, and Exponential Linear Units (ELU) [7, 211]. The activation function allows the MLP algorithm to fit non-linear functions with a powerful expression capability. The MLP uses one-way directed interconnections between the input, hidden, and output layers without recurrence. The hidden layer involves one or more layers. MLP utilizes the supervised learning technique *backpropagation* for training [197, 199]. MLP algorithm have good fault tolerance because all quantitative or qualitative information is equally distributed and stored in neurons [222]. It can sufficiently approximate complicated non-linear relationships with associative memory function. MLP is a complex algorithm for predictive modeling because several configuration parameters (e.g., threshold, learning rate, and momentum constant) need to be tuned, some based on automated support in WEKA [80, 181, 226].

5.3.2   Processing Machine Learning Training Data

The predictive module in this sprint feedback concept builds upon distinct supervised learning algorithms (see Section 2.2.4) using a unified (shared) training data format for training the ML models. The shared data allows a comparable validation of the individually resulting prediction performances of each model. The well-known attribute-relation file format (ARFF) is used as a unified training data format for building and validating the ML models under consideration [38]. Therefore, an automated processing routine transforms the socio-technical data foundation in Jira weekly into the ML suitable format. The ARFF separated header information and the respective data values, makes it an easily applicable input source for many supervised learning algorithms. The training data for the ML algorithms originates from the preprocessed socio-technical data foundation, which was weekly transformed into the ARFF scheme, as shown in Figure 5.12.

1. **Header Section** provides the name list of the included socio-technical attributes and the respective data types. The last attribute in the list refers to the ML model's targeted (prediction) attribute.

2. **Data Section** contains a list of the weekly socio-technical data objects. Each row comprises a data object covering numeric data values of the included attributes by week and team.

Figure 5.12: Schema for the ML Model Training Data

The ARFF contains the defined independent variables (explanatory) and the dependent variable (prediction) to build the ML model. A trained ML model can predict the dependent variable based on the relationship to the explanatory variables. Consequently, multiple data sets, each with a different socio-technical metric as the dependent variable, are needed for weekly ML predictions of all socio-technical metrics using differently trained ML models based on the same algorithm.

5.3.3 Performance-Based Selection of Prediction Model

In this predictive module concept, the previously described training data set is accessed in a subsequent performance-based model selection routine to determine the best-performing machine learning algorithm out of five different types. A shared training data set for different algorithms allows comparing the resulting prediction performances of the trained models. Moreover, it allows for the selection of the most accurate model type that best interprets the teams' unique development behaviors based on socio-technical dependencies in sprints.

In this context, an automatic feature selection (i.e., wrapper-based feature selection method) is performed on the training data set before the actual training of the ML models [221]. The additionally applied feature selection ideally reduces the number of explanatory attributes (independent variables), resulting in an insufficient relationship to the target attribute (dependent variable). It reduces possible input parameters for the derived regression-based prediction function, enhancing the accuracy due to the less irrelevant parameters.

Feature Selection for Model-Optimized Training Data

An integrated feature selection method supports filtering the independent variables (explanatory) within the ML-modeling process to only meaningful and significant relationships to the dependent variable (target/predicted). In this predictive module concept, a wrapper-based feature selection method was implemented to automatically search and filter the shared socio-technical training data to only significant attributes before the ML-model training. The AFS was individually applied to each of the five included ML algorithms needed for model-specific performance validation of varying feature subsets. The AFS process is shown in Figure 5.13.

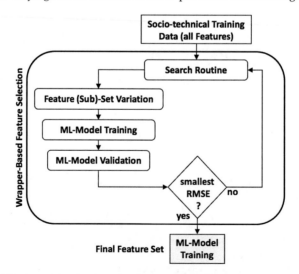

Figure 5.13: Feature Selection Routine Prior Training the Final ML Model, cf. [221]

The wrapper-based feature selection method searches for the most suitable feature (sub)-set (i.e., the combination of independent variables to explain the dependent variable functionally). Feature subsets are iteratively formed, whereby each ML model is trained and the prediction performance is determined using cross-validation (see Section 2.2.4).

The feature set that resulted in the lowest root mean square error (RMSE) is used for training the final ML model. In ML, a commonly applied measure for the effectiveness of a prediction model is the mean squared error (MSE), which is defined as the squared difference between the predicted value of parameter $\hat{\Theta}$ and the actual value of the population parameter Θ and defined as follows [66]:

$$MSE(\Theta) = E(\hat{\Theta} - \Theta)^2 \tag{5.1}$$

In this work, the RMSE is used to compare the prediction models' magnitude of the errors, defined as the square root of the MSE [66]. Due to the squared error, the resulting value is the mean-variance for even highly inaccurate predictions, and always positive [148]. The RMSE indicates the prediction model's absolute measure of fit to the data, whereas the well-known R^2 is only the fit's relative measure. Consequently, a lower RMSE value indicates a better data fitting model.

Model Training and Team-Based Performance Validation

The weekly training of the distinct ML models and subsequent validation of the prediction accuracy of each model is of central relevance for this concept ensuring the optimal weekly-updating selection of the most accurate algorithm corresponding to a team's socio-technical dependencies. Secion 5.3.2 described the processing of the ML training data results in an ARFF, covering the socio-technical data of all projects managed in a Jira system, which is directly applicable to all five ML algorithms. The subsequently applied feature-selection routine (model-bound) leads to a reduced training data set optimized based on each algorithm's preferable training data foundation.

These filtered training data foundation for each of the five ML models is used to train and cross-validate the prediction performance following the standard supervised learning process described in Section 2.2.4. The weekly training performance validation is performed for every (project) team individually towards identifying one of the five ML models' that fits best (by means of most accurate predictions) in response to a team's unique development behavior every week. Figure 5.14 and 5.15 visually discloses the relevance for an individual team-dependent model selection instead of a generalized one. The training set for the ML-models is based on socio-technical data captured in 13 student software projects during an observational study that concerned the technological feasibility of the computer-aided sprint feedback [134, 139]. The total project durations were 15 weeks, whereas the active usage of Jira was shortened by one preparation week, two weeks of holiday, and the last week was primarily used to create a product presentation.

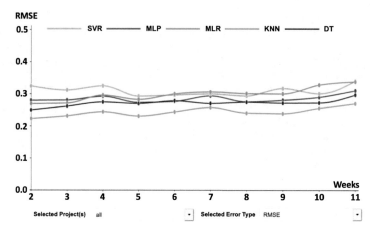

Figure 5.14: Weekly RMSE from Different ML-Models of Thirteen Student Projects

Figure 5.14 summarize the weekly differing RMSE of the five ML algorithms over time resulting from the predicted values of the thirteen projects' socio-technical data features compared to the actual captured metric values. The KNN achieves a weekly lowest RMSE with an average of 0.24 of maximal 1.0 and would generally derive steady predictions considering all projects' data. The result is understandable since the student software projects are somehow very similar to one another. The ML model comparison also confirms that the often applied linear prediction model would have fitted the data at worst, as depicted by the overall high RMSE of the MLR. However, using a single ML prediction model can put other projects at a disadvantage in terms of predictability. Figure 5.15 shows an example of a single student software project depicting the RMSE fluctuation of different ML algorithms for interpreting the team's socio-technical dependencies over time.

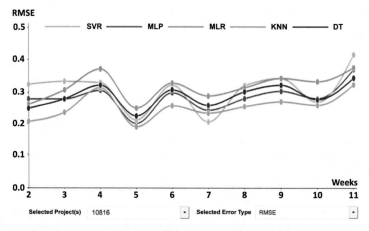

Figure 5.15: Weekly RMSE from Different ML-Models of Sample Student Project

The line chart in Figure 5.15 characterizes the changing RMSE of the five ML algorithms for weekly cross-validated predictions from only one of the student software projects (randomly selected). The RMSE of all five ML models showed an overall higher fluctuation but with almost similar courses over time. The KNN performed best for most of the weeks, except in the fourth week, where the MLP accomplished a slightly lower RMSE, and in the seventh week, the SVM performed significantly better than all other models. Nevertheless, the KNN could interpret the sample project's socio-technical behavior well on average.

However, the RMSE deviations have shown that a project-dependent ML model selection concerning the team behavior characteristics and socio-technical changes is essential. That is especially the case for software projects that would be less comparable considering the student software projects (more validation results of an industrial case study are described in Chapter 6). The automated model selection (AMS) routine has been established for the weekly team behavior predictions based on their distinct socio-technical characteristics and changes over time, thus selection of the machine learning model that interprets the data dependencies best.

Automated Model Selection Routine

This predictive module in this chapter covers an ensemble of five ML algorithms based on their distinct strengths and weaknesses for predicting socio-technical aspects in agile development teams (see process overview in Figure 5.11). The weekly processing routine involves automated feature selection for each ML model based on the preprocessed socio-technical training data, followed by subsequent training and performance cross-validation to determine the regression-based prediction accomplished by week and by the project (based on the RMSE).

The automated model selection (AMS) is the last processing step in the module's prediction support, which builds on each project's weekly RMSE outcomes to dynamically select the best-fitting ML algorithm with the lowest average RMSE over the previous three weeks. Subsequently, the selected ML algorithm is used for the ML training and prediction of the following week's socio-technical features based on a standard supervised learning process (see Section 2.2.4). Thus, the performance-based model selection provides weekly predictions that extend the linear regression models often applied alone and even automatically changes to the ML algorithm that fits best to interpret the recent trend of the socio-technical dependencies. The performance-based selection process for the most optimal prediction model consists of certain analogies from the well-known AutoML [73, 105, 226].

Based on the AMS routine, the overall prediction performance of the student software projects resulted in an overall RMSE improvement (e.g., 5.6% compared to the static use of the KNN and 8.2% compared to the static use of linear regression models). Moreover, the automated routine reduces a teams' manual effort to zero, enabling them to benefit from computer-aided sprint indications without hampering the daily development tasks. In this context, a visualization concept for the predictive module is realized in Jira to practically support teams with visual information that conforms with the previously introduced retrospective feedback.

5.3.4   Visualization Concept of the Predictive Module

The data processing of the predictive module covers the automated training and selection of the most suitable ML model for predicting the socio-technical aspects in this work. The computerized routine is executed weekly in the backend of a Jira system. Afterward, the module-related data is accessible in Jira's web frontend through encrypted REST calls. An additional visualization concept for the predictive module has been developed to visualize pre-analyzed future tendencies within a foreseeable time range (two project weeks) to support teams in sprints. Given the socio-technical aspects captured, the visualization concept follows the previously defined GQM goals for understanding team dynamics during sprints. The visualization concept of the predictive module is supplementary to the retrospective sprint feedback.

Line Chart Visualization of Team Behavior Courses in the Past and Future

The visualization concept provides a timeline chart summarizing teams' socio-technical aspects for the previous ten weeks and an estimated preview for the following two project weeks (based on the best-fitting ML algorithm listed in the bottom-left corner of the chart). Figure 5.16 shows the visualization of the predictive module, which is prototypically realized for projects managed in Jira. The example uses data from one of the observed student software projects.

A multi-scale (y-axis) enables the team to simultaneously compare the course of up to ten different socio-technical aspects without distorting significant course changes related to the time (e.g., based on unequal data types and value ranges). It characterizes the courses of the three selected sample attributes: *Communication – Flow Distance, Meeting – Quantity,* and *Mood – Positive Affect.* The course of the past ten weeks provides insights into the predicted course for the next two weeks and enables the teams to assess its plausibility. In the example chart, the automatically predicted course of the weekly alternating *Meeting – Quantity* seems to fit ideally with the previous course of the project over the previous ten weeks, and the *Communicatin – Flow Distance* and *Mood – Positive Affect* also follow a reasonable course. A line-hover function supplies weekly details about an attribute's intermediate course compared to a general trend (e.g., a slightly more positive mood than the average over the previous ten weeks). Descriptive measures explain the variance in the weekly socio-technical data (1st Quartile, Median, and 3rd Quartile), as in the retrospective modules.

The predictive module underlines the potential for using automated machine learning processing to predict future trends in complex socio-technical dependencies and the team behaviors resulting from them. However, the model processing described and the visualization possibilities also show that the ML models cannot provide humanly interpretable dependencies or actionable insights (except for the decision layers of a DT, which are roughly comparable over time). This thesis completes the computer-aided sprint feedback support with a final exploratory module covering visualized sprint dependencies as network graphs and a sprint planning cockpit built upon an adjustable simulation model.

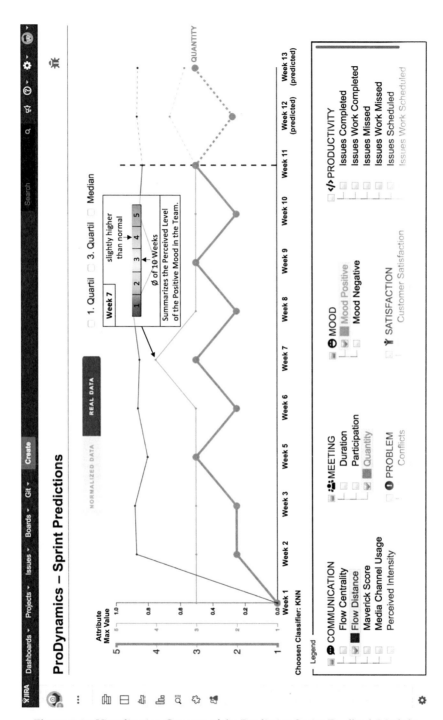

Figure 5.16: Visualization Concept of the Predictive Sprint Feedback Module

5.4   Exploratory Sprint Feedback Asset

The previous sections described the descriptive and predictive feedback assets that support awareness and understanding of past team behavior trends and deviations as well as ML-based estimations as supplementary sprint feedback in ASD for the immediate future. The objectives were to provide low-effort and time-saving support that enables team members to practically track socio-technical aspects and gain supportive insights during the development. Both feedback modules provide greater information accessibility and insights into past and future sprint performances based on systematic capturing and automated processing of team-related socio-technical aspects.

However, the previous modules focus only on how development performances change over time rather than providing feedback on why they change (a common problem in statistical learning). Furthermore, deriving actionable implications based on socio-technical dependencies in sprints can be challenging to learn and understood by the human mind, compared to machine learning [128, 134]. In this context, agile practices magnify the impacts of development dynamics, such as iterative customer feedback, interventions based on change requests, and the relationship between schedule pressure and estimated employee capacities, which can reflect sophisticated dependencies. Successful ASD requires an understanding of arising dynamics that can lead to advantageous action-taking in ongoing or future projects [41].

The additional exploratory asset in this computer-aided sprint feedback concept covers two feedback modules to support teams' capacity to interpret socio-technical dependencies in sprints with the help of straightforward human-understandable visualizations and models. Both can be used to explore significant socio-technical dependencies in sprints (e.g., changed team communications, customer satisfaction concerning performances). Therefore, the information support in the first module involves network graphs, which support the exploration of visualized relationships between the socio-technical aspects. Exploratory data analyses are applied to characterize the sprint dependencies (see Chapter 2). At the same time, network graphs are used to support a comprehensive visualization of teams' interrelating socio-technical factors in a specific sprint or across multiple over time. The second module's information support relies on a system dynamics model that enables a simulation-based exploration of different sprint scenarios before the actual start of a new. The simulation model has been prototypically developed for Jira and includes an interactive interface for exploratory sprint planning, useful for teams without SD experience.

5.4.1   Sprint Dependencies Module

The extensive retrospective and predictive sprint feedback modules provide weekly support for agile teams to better understand the average courses, fluctuations, and detected anomalies of individual socio-technical aspects in sprints.

While the predictive module's ML algorithms support linear and non-linear regression-based predictions, the resulting models and prediction outcomes lack the simplicity necessary for teams to directly interpret the statistically analyzed socio-technical dependencies (e.g., how the selected explanatory variables relate to the predicted). In the introduction of this thesis, Ackoff's [6] DIKW pyramid (see Section 1.2.1) was used to emphasize that information technologies can support the characterization and mining of meaningful information (e.g., patterns and relationships in data) and that people have to be able to subsequently understand and internalize the processed information before taking actionable decisions. The objective of this sprint dependency module is to provide computer-aided feedback support to promote cognitive understanding of how each socio-technical aspect relates to another in past sprints. Figure 5.17 summarizes the sprint dependency module's automated processing routine.

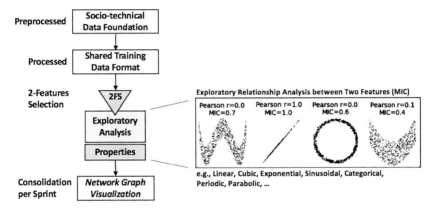

Figure 5.17: Automated Dependency Analyses and Visualization Process

This dependency module primarily covers an automated two-step processing routine in every sprint that performs non-parametric data analysis for exploratory characterization of the relationship between two selected socio-technical features. It derives visualized network graphs for the significant identified dependencies [134, 194]. The exploratory analysis routine uses the same shared ML training data set as previously described for the predictive module (see Section 5.3.2).

5.4.1.1 Automated Exploratory Analyses

The automated characterization of the socio-technical sprint dependencies builds on an exploratory algorithm that uses the maximal information coefficient (MIC) as a novel measure for identifying and classifying even complex associations between pairs of data variables (see Section 2.2.5) [134, 194]. The MIC is a relationship coefficient score based on multiple statistical correlation measures for identifying dependencies in up to 27 different functional relationship types (e.g., the Pearson correlation coefficient, Spearman's rank correlation coefficient, mutual information estimation, maximal correlation estimation, and curve-based dependency measures)

[58, 141, 194]. Consequently, the algorithm can systematically characterize the functional properties of data relationships (e.g., simple linear and more complex functional dependence). The purpose of including the automated exploratory analysis is to ensure that all functional relationship properties of the data pairs are adequately investigated and that no significant relationship is overlooked or underestimated, which might occur with only a linearity analysis [137, 194]. Data complexity, a lack of time for manual processing, or a lack of analytical expertise should not present obstacles to the accurate interpretation of dependencies between socio-technical aspects in projects.

The exploratory analysis in this concept is automatically performed at the end of every sprint. The processing routine uses the weekly updated ML training data set from the predictive module and filters the available data entries (for all socio-technical features) according to the active period of the sprint to be analyzed. Subsequently, $\binom{n}{k}$-possible data subsets for all two-feature combinations without repetition are derived, where k is the number of selected features from the total number of n features. The MIC algorithm analyses the functional relationship properties of every 2-feature combination individually based on their mutual information. The algorithm is adapted for this concept through sprint-based preprocessing of the socio-technical input data. This involves computational performance optimization using multithread support, which allows for parallel analyses of the feature pairs and an adjusted output format that supports the visualization of network graphs. Table 5.2 shows an example of the algorithm's output scheme (ranked list), which applies to all analyzed two-feature relationship findings.

Table 5.2: Example of the Exploratory Analysis Output Format

Feature X	Feature Y	MIC	P-Value	R-Type	Polarity
Meeting Duration	*Meeting Participation*	*1.00*	*0.02*	*linear*	*negative*
Media Channel Usage	*Meeting Duration*	*1.00*	*0.03*	*non-linear*	*positive*
Maverick Score	*Meeting Participation*	*0.94*	*0.04*	*linear*	*negative*
.	.	.	.		.
.	.	.	.		.
.	.	.	.		.

The non-significant relationship findings with a p-Value > .05 are automatically filtered during the exploratory processing and are, thus, not considered in the ranked list. Inclusion would mean that non-significant relationships are considered for the network graphs, which would hamper the cognitive interpretability and visualization of only the relevant socio-technical dependencies in sprints. Besides the automated processing of the exploratory data analyses, the sprint-related dependency findings are made accessible through a REST call in the frontend of Jira used by the prototypically realized visualization concept of the sprint dependency module. The automated processing has evolved from solely manual analyses to semi-automated application support with network graph visualization, which has led to this concept's completely automated sprint dependency module [35, 134, 137].

5.4.1.2   Visualizing Sprint Dependencies Through Network Graphs

Network graphs based on dependencies are a simple visualization form for exploring and depicting interrelated socio-technical aspects, allowing team members to recognize the relevant changes more quickly and efficiently in sprints. Moreover, the automated processing in this concept can help managers and teams to understand insights into significant dependencies in previous sprints without any manual effort or expert knowledge on how to characterize the sometimes complex data dependencies [134].

Like many other visualization techniques, network graphs can promote cognitive understanding and awareness of interrelations between different entities (nodes) and the connections (edges) that characterize relationship strengths (e.g., weighted edges). The visualization concept for the analyzed socio-technical dependencies in sprints builds on the outcome of the exploratory analysis using the adapted MIC algorithm. The dependencies are visualized through eight graph components, each containing different information properties. Each of these components provides additional characteristics accessible through visual and interactive elements (e.g., through the selection or hover functions of the graph components). Table 5.3 lists the network graph components used to model the dependencies.

Table 5.3: Network Graph Components for Visualizing Sprint Dependencies

Graph Component		Description
ⓧ⁓Ⓨ	Node-Source	The data object X in the socio-technical relationship.
ⓧ⁓Ⓨ	Node-Target	The data object Y in the socio-technical relationship.
●◐◯	Node-Group	The node-color bases on the information category.
⊘	Node-Size	The diameter bases on all connected edge-weights.
≡	Edge-Width	The dependency strength between Node X and Y.
+/–	Edge-Polarity	Indicates a positive or negative relationship affect.
ⓘ	Edge-Hint	Textual interpretation of the relationship properties.
ⓧⁱ	Node-Hint	Textual interpretation of the Node properties.

The theoretical foundation for visualizing the network graphs based on sprint dependencies originates from the previously described team communication network graphs and social networks (see Sections 5.2.2.2 and 2.1.4). Every significant relationship between a pair of socio-technical features (nodes) is visualized through a weighted edge that connects the two entities. The color of the graph nodes depends on its predefined association with a provided information category (e.g., the communication intensity, media channel usage, or the Flow Distance relate to the Communication). The analyzed MIC scores describe the relationship strength and are depicted through auto-scaled edge widths.

In this context, the dependency characteristics analyzed by the MIC algorithm do not distinguish between different relation directions (i.e., between Node-X to Node-Y and Node-Y to Node-X). Consequently, all dependencies are visualized using undirected edges. These diameter auto-scales are based on the sum of all absolute values of in- and outgoing edge-weights. The different node sizes imply the relevance of a sole socio-technical aspect during a sprint. The node-hint covers a tooltip that summarizes some of the selected node's characteristics (i.e., the information category, a short description of the feature, a dependence rank compared to the other features, and the total number of relationships).

Every edge covers a supplementary polarity mark and a (generic) textual hint depicting the analyzed relationship properties when hovered over. The polarity mark between two socio-technical features indicates whether the relationship is positive or negative (i.e., as the value of Node-X increases over time, that of Node-Y also increases, and vice versa). In a negative relationship, a similar analogy applies (i.e., the value of Node-X increases over time while the value of Node-Y decreases and vice versa). An edge-hint holds a tooltip that summarizes the characteristics of the selected relationship (i.e., linear or non-linear dependency type and the strength corresponding to the analyzed MIC score). The supplementary information supports a simple exploration of interrelating socio-technical aspects in sprints, which cannot be expressed as well by numeric results alone (e.g., ranked lists or data tables). Figure 5.18 shows an example of the visualization concept in the Jira environment with a sub-graph of socio-technical dependencies from one of the observed student software projects.

Extended Details and Sub-Graphs

1. *Information Category* - The selection of socio-technical information categories shows all relationships identified in the selected category. A category tooltip depicts all associated aspects in this group, which are considered during the exploratory analysis process.

2. *Single Node* - The selection of a particular socio-technical aspect filters out all non-related dependencies. A node tooltip summarizes the relevant properties of the aspect.

3. *Pair-of-Nodes* - The selection of a specific relationship edge filters out from the network graph all relationships not selected. An edge tooltip summarizes the analyzed relationship properties of the two-feature dependency.

4. *Relationship Strength* - The relationship severity scale supports three selected options for filtering all relationships below the selected strength. The scale was adapted from a conventional correlation coefficient binning [223].

5. *Sprint* - The sprint selection updates the socio-technical dependencies graph based on the relationship findings analyzed previously in the backend routine. The "Across-View" checkbox enables a visualization of the network graph for all socio-technical dependencies of the completed sprint up to the selected sprint.

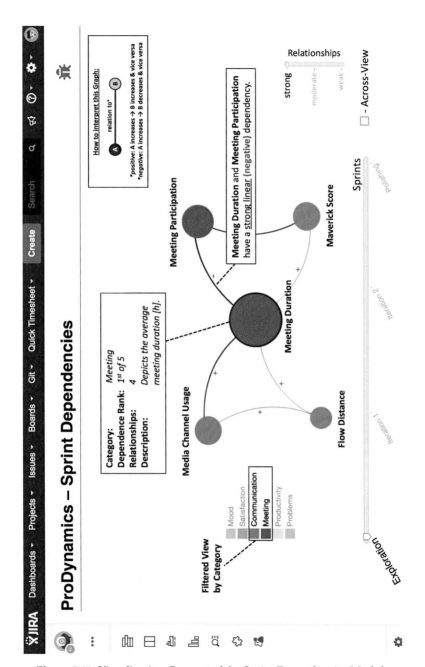

Figure 5.18: Visualization Concept of the Sprint Dependencies Module

The network graph provides interactive elements to channel the dependencies for concrete representation based on filtered sub-graphs. The visualization in a network graph supports filtering by relationship strength, category selections, specific relationship selections, or specific node selections, which filter the view to related aspects only. The interactive functions of the graph components (e.g., tooltips and filters) provide additional insights into the relationship properties. In the example shown in Figure 5.18, a specific node (Meeting Duration) is selected to filter out all socio-technical aspects with no relationship to the selection. An additional selection of the category filter (Communication and Meeting) then reduces the view to only those aspects that are associated with the two categories.

The concept supports the exploration of interrelated socio-technical aspects in and between past sprints. Moreover, the visualization of socio-technical relationships over time (across multiple sprints) is subsequently available through an across-view option. The visualization of relationships as a network graph supports cognitive perceptions useful for enabling teams and leaders to investigate problems and plan the next sprint (e.g., communication changes and past dependent aspects) [134].

Interpretation of the Sprint Dependencies Network Graph

The visualization in Figure 5.18 is based on sample data from the observational study that involved student software projects aimed at obtaining a better understanding of the team dynamics in sprints [134]. The selection of the single node Meeting Duration results in a sub-graph based on a filtered view covering only the dependencies of the selected element. The sub-graph is especially useful for highlighting only relevant aspects of large or sophisticated network graphs. The directly dependent socio-technical aspects Flow Distance and Media Channel Usage showed an additional indirect relationship that revealed a reinforcing relationship cycle with the selected node.

The interpretation of the network graph for the selected sprint is as follows: As the Meeting Duration increased, the Media Channel Usage and the Flow Distance increased as well, and the Meeting Duration decreased when Media Channel Usage was lower. Furthermore, the Meeting Participation ratio decreased when the meetings lasted longer. Thus, long team meetings resulted in significantly more maverick activities in the team. At the same time, whenever maverick activities increased, meeting participation rates dropped.

Important Remark: The network graph's dependencies reflect only the subjectively and objectively captured development behaviors of a single team for a particular period. External influences (e.g., project environments) can change over time, as can groups. The sprint dependencies module visually depicts significant relationships within a realistic range, but these relationships are limited to the socio-technical aspects considered in this work. Nevertheless, the concept supports the recognition and understanding of these aspects in sprints.

5.4.2 Sprint Dynamics Module

The previously described sprint dependency module enables feedback for development teams covering the static visualization of socio-technical dependencies arisen during each past sprints, automatically processed through exploratory data relationship analyses (see Section 5.4.1). Team dynamics encompass "soft factors" and effects over time that influence the development performance of teams [81]. This supplementary sprint dynamics module supports the awareness in projects concerning team dynamics through a simulation-based exploration of socio-technical relationships in sprints.

The concept of this module builds on methods for data inclusion in System Dynamics Models (SDM) [102]. Data methods (e.g., exploratory data analysis, machine learning, and statistical screening) are combined with mental data (e.g., expert knowledge, interviews, experience-based relationship assumptions) to exceed the limitation of data availability. The incorporation results in a sprint simulation model that supports exploring different sprint scenarios based on adjustable parameter settings (e.g., to simulate possible development bottlenecks based on reduced developer capacities or productivity in a future sprint). The simulation model was built for the context of agile software development and evaluated in an industrial pilot study [128]. The following modeling process for an exploratory sprint planning support led to the sprint dynamics module for Jira.

5.4.2.1 Data Inclusion for System Dynamics Modeling

Building the system dynamics model required understanding the dependencies of the target system (i.e., socio-technical aspects related to development performances in sprints). Therefore, process-related knowledge was essential to understand relevant sprint dependencies (endogenous and exogenous components) throughout the four stages of the modeling process described by Sterman and Albin [8, 216]:

1. **Model Conceptualization** - System Boundary and Dependencies

2. **Model Formulation** - Parametrization and Relationship Equations

3. **Model Testing** - Verification of Model Abilities

4. **Model Evaluation** - Simulation Inferences for the Real-World

In this context, the data methods included in this study for the modeling, such as exploratory data analysis, visualization techniques, FLOW interviews, support the intuitive understanding of sprint dependencies based on past data incorporated with experiences, observations, and assumptions of practitioners about team behavior [102, 127, 128]. Figure 5.19 summarizes the adapted modeling process in a collaborative study between Arvato SCM Solutions and the Leibniz University Hannover to build the sprint dynamics model [128]

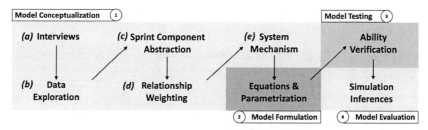

Figure 5.19: Sprint Dynamics Modeling Process

Model Conceptualization

Identifying the system boundary and the mechanism for modeling sprint dynamics takes data analysis methods and practitioner expertise into account to discloses experiences with dependencies that cannot be explained by project data alone.

(a) Interviews: The FLOW method [208] is a well-established method in software engineering used to systematically capture the perception-based information and process structures within teams, projects, and the entire organization. In this context, FLOW interviews with the different project roles at Arvato disclosed insights into internal process structures and dependencies in industrial agile development environments relevant for the model conceptualization. A template for the FLOW interviews structures the conversation through a set of general questions, such as the tasks within the team or the experience of the interviewees (see Appendix B.1). The consolidation of the individual process-related responses resulted in a fundamental understanding of the Scrum teams' development processes [128].

(b) Data Exploration: Process data available from previous projects or sprints can help to substantiate the qualitative interview responses quantitively or derive additional insights for sprint dependencies less perceptual for the human mind. Moreover, available data is helpful to test the simulation model's abilities in the end. For the data exploration, past project data in Jira from three teams about 22 sprints (220 workdays) has been accessible. The sprint data covered the number of the change requests, estimated vs. completed story points, estimated versus required work hours, and velocity in each sprint. In addition, the management manually elicited customer satisfaction at the end of sprints on a five-point scale. Other sociological data (see Chapter 4) was not available either.

(c) Sprint Component Abstraction: As the starting foundation for the continuing model conceptualization, a first sprint causality diagram was derived considering the quantitative data and interviews responses. Figure 5.20 shows the causal diagram on a high abstraction level of assumed and intuitive relationships involving central aspects like *HR-capacity*, *Mood*, and *Customer Satisfaction*. The availability of developers (HR capacity) has a perceived impact on the commitment constancy (i.e., the difference between estimated story points and completed in a sprint), simultaneous to the sprint progress.

At the same time, the achieved commitment constancy at the sprint end influences customer satisfaction with an effect on the mood in teams. The mood, in turn, is perceived relevant for the sprint progress, thus reflecting the status in the burn-down chart.

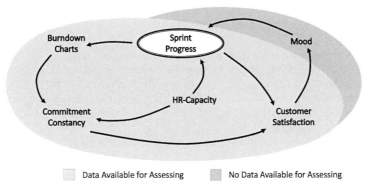

Figure 5.20: Preliminary Sprint Relationship Diagram

In particular, the interviews disclosed the relevance of sprint progress for achieving a commitment constancy within the target range (±15% of the sprint estimations concerning the story points and HR capacity). Consequently, the burndown charts are crucial information assets for tracking the sprint progress and targets.

(d) Relationship Weighting:

A consecutive workshop with seven participants from Arvato was conducted to identify possible further relationships and model components extending the causal diagram abstracted in Figure 5.20 [128]. The workshop goal included adding/removing system components, establishing connections, and determining the direction of influence. In addition, all participants were asked to rate the expected relevance of the relationships using a scale ranging from -3 to +3. A positive relationship influence is indicated by (+), whereas a negative is marked by a (-) beside the arrowhead. The numbers indicate the assumed relevance (1=weak, 2=medium, 3=strong relationship). The joint group assessment led to a common understanding of the model dependencies with relevance for identifying the next steps of system mechanism in the conceptualization process. An excerpt of the intermediate causality rating during the workshop is attached in Appendix B1.4. The ratings disclosed the assumed relevance of relationships that otherwise could not be detected due to a lack of data (e.g., the influence of customer satisfaction on team mood). However, the relevance ratings do not characterize the relationships.

(e) System Mechanism: The concretion of the system mechanisms is an iterative task in which the interrelating components require conversion into a static stock-flow diagram. The conversion has the goal (and challenge) to make the abstract component from the causal diagram quantitatively explainable without quantifying them yet. The identification of explainable metrics and variables (auxiliaries) in this work

is done based on the discussion outcomes during the workshop and the past project data. It is an exploratory activity seeking to explain specific dependencies. For example, the sprint progress in a simple form is describable solely by objective measures like the development productivity over time and is an easy starting point for the modeling. Besides, productivity also has an intuitive dependency on HR capacity, burndown charts, commitment constancy, and mood in a team without clearly knowing the detail of the relationship.

Another critical factor is the time granularity of the targeted simulation model. The sprint dynamics module aims to support the planning activities. Consequently, in this example, the considered metrics should explain productivity on a daily basis, allowing to characterize productivity differences for regular weekdays, holidays, and events (customer audits). Figure 5.21 shows how the initial sprint progress component changed into productivity connected to the commitment constancy through a burndown chart module that covers successively added explainable auxiliaries, stocks, and flows of the relationship.

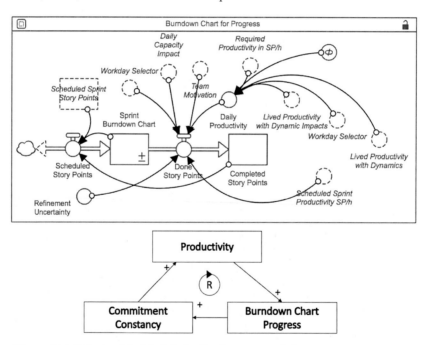

Figure 5.21: Relationship Module for Productivity and Commitment Constancy

Modeling of relationships is not trivial and requires iterative adjustments for newly added elements. Thus models fastly increase in complexity that is hard to overlook [14]. Current versions of system dynamics modeling frameworks support modular structuring through interfaces (dashed objects). The sprint dynamics model of this work emerged during the case study with Arvato [128]. However, it is essential to understand that the modeling and formulation are also iterative, hand-in-hand activities for every newly added component.

Model Formulation

The model formulation is the second modeling task turning a static stock-flow model into a system dynamics model through relationship equations and parametrization. The mathematical equations are defined based on assumptions (logical associations) or knowledge about the relationship properties. For example, the following equation formulates the stock "Sprint Burndown Chart" shown before in Figure 5.21.

INIT *Sprint_Burndown_Chart = 0*
UNITS:
Story Points

INFLOWS:
Scheduled_Story_Points =
Scheduled_Sprint_Story_Points - Sprint_Burndown_Chart - Completed_Story_Points
UNITS:
Story Points/Day

OUTFLOWS:
Done_Story_Points = Workday_Selector * Team_Motivation*Daily_Productivity +
("Scheduled_Sprint_Productivity_SP/h" * Refinement_Uncertainty) * Daily_Capacity_Impact
UNITS: *Story Points/Day*

Nowadays, system dynamics modeling frameworks also offer graph-based equation support, as shown in Figure 5.22. These dynamic equations were used in two cases in this model: (1) exploratory data analyzes characterized a significant relationship or (2) a data-driven explanation was not possible, so equation assumptions (e.g., experience and learning curves in a team) were used to formulate the dynamics of the component.

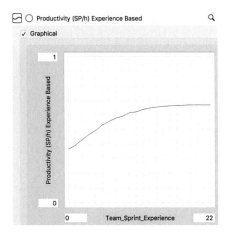

Figure 5.22: Graph-Based Equation Example for Dynamic Auxiliary

An advantage of graph-based equations is that they are easily adjustable to explore relationship characteristics to optimize the model. In addition, for dependent components without available objective data records, the functional equations involved the relevance ratings for the assumed dependencies (similar to regression coefficients in a normalized function).

Model Testing

The model testing is the third central activity during the system dynamics modeling process. Model tests can help validate whether the model's simulation ability matches the expectations, or better, actual reference data for comparison. In previous work, sensitivity analyses were applied first for verifying the sprint dynamics model's correct functional operating based on realistic input ranges. A precondition for the sensitivity analysis concerned the two most relevant sprint planning aspects: HR capacity, and story points. While other input ranges may be realistic, the value ranges for human resources (60-150 working hours) and story points (80-120) were limited in this study. The sensitivity analyses covered a total of 9000 simulation runs with randomly sampled input parameter settings [128].

The model passed the first functional plausibility test regarding that all simulation outcomes resulted within realistic ranges. Some outputs extremes required additional confirmation for plausibility based on practitioner experiences. Moreover, the sensitivity analyses supported determining several behavior patterns in the model. Figure 5.23 shows a simulation excerpt regarding the resulted productivity course changes throughout sprints.

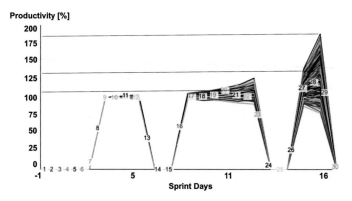

Figure 5.23: Relative Productivity Variance for 100 Sprint Samples, cf [128]

The chart shows the daily productivity (rational scale) course throughout a sprint. Days with productivity of 0% represent weekends. The most relevant finding due to parametric input variations was the course changes of the simulated productivity, which disclosed an almost constant outcome during the first weeks in a newly started sprint. In contrast, the simulation results revealed that more dynamics in the sprints occurred during the second week - with a slight overall tendency for higher productivity than in the first week.

Remarkably, particularly the last three days of a sprint revealed a noticeable change in daily productivity. The analyses were performed for a regular sprint with ten days without event impacts, such as holiday breaks. The simulations pointed out that the teams typically seem to finish story cards with an overall productivity increase of 32%, and in extreme cases, they even double speed compared to the productivity note during the first sprint week.

Besides, the productivity increased throughout a sprint and, in some cases, almost doubled relative to the productivity at sprint start. Nevertheless, the productivity increases can also imply that teams started with reduced performance and then increased their regular performance shortly before the due date to keep up the sprint goals. The project manager's perception was involved because such a behavioral explanation is limited to its subjective interpretability. He affirmed that his impression is that the teams tend to underperform at the beginning and increase their productivity near the end to fulfill the sprint goals.

Model Evaluation

Evaluating the system dynamics model concerning simulation inferences in practice presents the last step of the modeling process. If possible, as in this work with Arvato, a workshop with practitioners supports jointly exploring the model's sprint planning credibility through different user perspectives and scenarios. One must keep in mind that the model is not forecasting exact values, but to disclose the dynamics of sprint dependencies for more awareness during the planning.

The participating developers, PO, and a manager tested the different functional modules for their input and output behavior (see Appendix B2.1) regarding different sprint conditions they had in mind (e.g., real scenarios from the past and synthetic ones). The scenarios from the past included ideal sprints courses without specific incidents to be compared with regularly satisfying sprint achievements, such as past commitment constancy and customer satisfaction. Non-ideal sprint scenarios evaluated the model's ability to manage incidents. The test scenarios covered, for example, increased sprint impacts concerning changed story points and fluctuations in the capacity.

The simulation findings disclosed the managements initially perceived underperformance of the teams at sprint start. Moreover, some sprint courses led to "wow"-effects among the participants because of non-expected but plausible outcomes. For example, the course of mood in the simulated sprints presented another affirmative observation that represented the close reality, given the perception of the product owner and the developers. The observation is valuable because it confirms the proper formalization of the mood dependencies, which build upon assumption during the first model conceptualization workshop at Arvato.

The system dynamics model simulates the sprint dynamics corresponding to the practitioners from the three teams at Arvato and behaves intuitively in many natural and synthetic sprint planning scenarios. The model simulations matched the overall expected (based on the simulation input) outcome and simulated results.

5.4.2.2 Visualization Concept of the Sprint Dynamics Module

The modular structure of the sprint dynamics model reduces the visual complexity due to less viewable links using component interfaces. However, the stock-flow charts might not be trivial to understand or even utilize by non-experts, given the many possible parameter settings. Therefore, a complementary dashboard visualization was built for the sprint dynamics model, enabling non-experts to perform simulation-based sprint planning without needing domain expertise or knowing all model dependencies. The dashboard was used in the model evaluation described before that allowed practitioners from Arvato to explore sprint dynamics based on varying planning scenarios intuitively. Figure 5.24 shows the dashboard, that was afterward the evaluation migrated to the ProDynamics plugin for Jira.

The dashboard is optimized for sprints with up three weeks in length. While a different parameter setting in the underlying system dynamics model also supports longer sprints, the dashboard limits this (model) setting to a maximum of 21 workdays (relevant for the start and end time of the simulations). Below the workday-select options are two input fields regarding the estimated HR capacity available and estimated story points for the next sprint. The information table on the right side of these two input fields summarizes a set of relevant sprint metrics, including

- the scheduled number of sprint days

- the scheduled productivity [SP/h] based on the planning

- the required productivity [SP/h] to complete sprint in time

- the total number of added or removed story points during a sprint

- the total number of added or removed HR capacity during a sprint

The simulated sprint course is visualized as a line chart, disclosing the several statuses of the daily updated parameter (e.g., expected commitment constancy if the team continues this way or remaining HR to keep the sprint ecological). The barometers show the daily status (relative to ideal course). Using the line chart allows a simple disclosure of whether sprints complete earlier, meet the scheduled target deadline, or remain incomplete at the sprint end (compared to the initial planning).

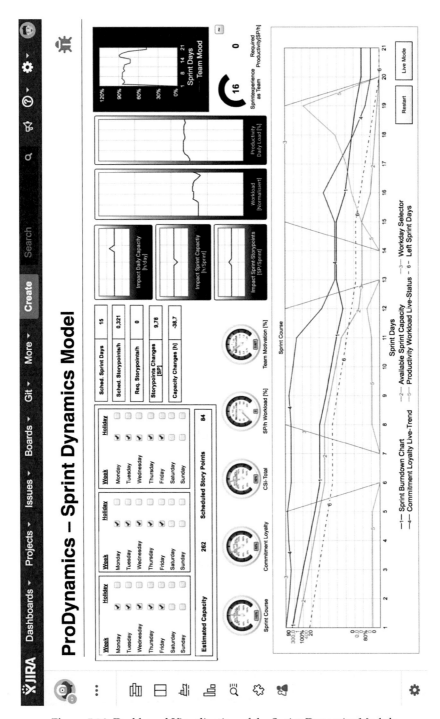

Figure 5.24: Dashboard Visualization of the Sprint Dynamics Module

5.5   Technological Feasibility of the Feedback Concept

The computer-aided sprint feedback concept extends the socio-technical data capture concept described in Chapter 4 through an automated descriptive, predictive, and exploratory characterization of the team behavior in agile software projects. The two concepts support the team's understanding of team behavior in ASD through complementary sprint feedback covering a systematic capture of socio-technical aspects. They also minimize the manual effort for complex data interpretation by involving automated processes and analytical methods from data science.

During this work, both concepts were iteratively developed between 2015 and 2020, following a design science research (DSR) methodology (see Section 1.3). This study identified relevant socio-technical information needs in ASD environments based on theoretical foundations, interviews, and observations. The consolidated knowledge foundation covers practical socio-technical metrics, measurement methods, and the data processing routines for the retrospective, predictive, and exploratory feedback assets about team behaviors [128, 134, 135, 137, 138, 139]. The primary goal of the DSR was to develop a technologically feasible concept for sprint feedback that supports understanding team dynamics. This goal was accomplished and refined through several independently conducted studies with 32 student software projects and three industry projects.

The preliminary findings from analyzing the student and industrial software projects affirmed the general technical feasibility of both conceptual practices. Concerning the three computer-aided sprints feedback assets, the observations provided insight into the socio-technical dependencies and, in turn, the team dynamics in the software projects. A final evaluation in unfettered industrial environments concerning the usability and utility of the consolidated computer-aided sprint feedback concept was applied using an action research approach involving two agile software projects, described further in Chapter 6. The following subsections summarize the significant observational findings during DSR that affirmed the technological feasibility of the three computer-aided sprint feedback assets and the underlying socio-technical data capture.

5.5.1   Descriptive and Predictive Feedback in Student Projects

The descriptive (retrospective) and predictive feedback assets concept that emerged during the DSR was approached for technological feasibility using a comparative study in 2018 involving 15 student software projects with 130 students [136, 139]. Each project was estimated for 15 weeks and involved four sprints (see Section 4.5). The study observed whether team behavior positively or negatively affects the organization and evaluated development performance when providing computer-aided sprint feedback throughout the projects. The Jira plugin, ProDynamics, was used to handle the systematic capture and analytical data processing of socio-technical aspects in sprints (e.g., weekly team communication and mood data from seven student teams interested in accessing additional feedback). Simultaneously, the sprint performance feedback by the customer and project manager (see Section 2.1.6) was observed through Jira for all teams, allowing for a comparison of team performances with and without ProDynamics.

The seven teams actively utilized the descriptive and predictive sprint feedback during the software projects (i.e., participation in weekly sociological surveys and access to the feedback modules). In contrast, the demand decreased in the last third of the projects' duration to only 1–2 active team members using ProDynamics. Nevertheless, considering that seven teams actively utilized the feedback and frequently contributed sociological information throughout several weeks, the descriptive sprint feedback asset is technologically applicable to student software projects. In addition, the captured customer and project manager feedback at the end of sprints without external efforts in collecting and processing the data fastened the availability of team feedback.

With this, 77% of all the members reported a perceived added value for their development outcomes when using the additional sprint feedback during the process [139]. Statistical analyses affirmed these positive perceived performance improvements for the seven teams that actively accessed the retrospective and predictive sprint feedback. These teams exhibited a more balanced velocity distribution with fewer estimation errors ($\pm 9\%$)than the other teams (± 19)%. The other eight teams without computer-aided sprint feedback had a strong tendency to overestimate their spring tasks during the first two sprints, followed by underestimations that caused substantial deviations between the number of scheduled tasks and completed ones [136].

5.5.2 Exploratory Feedback in Student and Industrial Projects

The exploratory feedback asset concept that emerged during the DSR resulted in sprint dependencies and dynamics modules. The sprint dependencies module was used for technological feasibility in 2019, involving six teams (out of 17 student software projects) with 53 students who used ProDynamics throughout their projects [134]. Each project was evaluated for 15 weeks and involved four sprints, similar to the prior technological feasibility approach concerning the descriptive and predictive feedback assets described in Section 4.5.

Considering the weekly sociological data contributed by the six teams, in extension to their development performances captured within Jira and performance satisfaction feedback from the customers and projects managers at the end of sprints, the teams yielded access to weekly and sprint-wise feedback on socio-technical sprint dependencies throughout the projects (see Section 5.4.1). The exploratory sprint dependencies, particularly their visualization concept through dependency network graphs, disclosed significant team behavior patterns. The dependency visualization helped the teams keep track of their development behaviors and systematically detected changes related to sprint performance variations.

Moreover, the sprint dependency characterizations in ProDynamics showed that all six teams saw a significant change in their development behavior for achieving high customer satisfaction, solely for the last sprint in the project [134]. In previous sprints, only the project managers' feedback was considered relevant, although the customers provided feedback throughout the projects. Even if someone in academic teaching considers this behavior trivial, it is astonishing to prove this quantitative using computer-aided methods.

The sprint dynamics module presents the second exploratory sprint feedback module. Its technological feasibility was part of a case study with three industrial software projects at Arvato SCM Solutions [128]. The underlying system dynamics model was conceptualized, formulated, validated, and subsequently evaluated in close cooperation with the practitioners of the involved projects (see Section 5.4.2). The system dynamics model was realized as an embedded web application that makes it cross-platform compatible, which is relevant during the conceptualization process for later utilization. The model supports the simulation-based exploration of plannable sprint scenarios for characterizing team performance variations, focusing on customer satisfaction, mood, and productivity influences.

The exploratory scenario planning enabled developers and managers to review past performances based on actual planning references and disclose possible sprint courses based on synthetic conditions (e.g., reduced capacity due to illness, varying workloads, or change requests events). The technological feasibility corresponded to the practitioners because the system dynamics model effectively simulated the reality and was easy to use through the sprint planning dashboard [128]. The evaluation of the sprint dynamics model at Arvato was based on a platform-independent web version[3] that was afterward adapted into the ProDynamics plugin for Jira to unite the retrospective, predictive, and exploratory feedback assets in a central sprint support platform concerning the socio-technical aspects in sprints.

---

[3]https://www.iseesystems.com

Chapter 6

Evaluation of the Computer-Aided Sprint Feedback Concept

Software engineering is a complex social activity, and the success or failure of processes, methods, and tools is highly dependent on the application context [185]. The previous two chapters of this thesis covered the concepts concerning the provision of computer-aided feedback to support an understanding of team behavior in sprints based on systematic capturing of socio-technical aspects and automated processing routines. The conceptualization followed a design science methodology for information system research involving interdisciplinary theories, metrics, and methods from social science, data science, and software engineering. The consolidated concept has been prototypically realized as an extension for Jira systems (referred to in this work as ProDynamics plugin).

The retrospective, predictive, and explorative feedback modules covered in the concept have been practically applied in observational studies and improved based on observations from 32 student software projects and three industry projects [128, 134, 139]. The studies demonstrated the technical feasibility of the computer-aided sprint feedback (see Section 5.5). Studies in student software project are often conducted first (more accessible environments) before being carried out in industrial projects [42].

This chapter presents an evaluation of the ProDynamics concept in an industrial case study with respect to its overall usability and utility based on practitioners' feedback. The study was conducted in accordance with an action research methodology, which facilitated the identification of the deficits of the concept and the development of improvement interventions, which were implemented in the ongoing projects in collaboration with the participating practitioners [183]. The evaluation covered cyclic qualitative interviews together with quantitative interpretations of the captured socio-technical aspects over time. Due to the complexity of the plugin, the computer-aided feedback support modules were sequentially assessed and improved (i.e., starting with the mechanisms for capturing socio-technical data and ending with the exploratory feedback module).

The assessments and improvement of the sprint feedback modules were conducted in sequential order to ensure that socio-technical data is accurately captured from the start of projects. Accurate data capture was essential since each sprint feedback support module builds on data from the previous retrospective, predictive, and exploratory modules. The action research has led to several practical improvements, which successively improved the usability of the plugin and, consequently, the utility of computer-aided sprint feedback in the context of industrial ASD. The following sections describe in detail the evaluation, including the quantitative and qualitative observations and conclusions.

6.1 Assessing ProDynamics in Industrial Software Projects

Experimental Software Engineering is essential to provide the foundations for understanding the limits and applicability of software technologies (e.g., the need to observe and pair best practices in Software Engineering with theoretical methods) [101]. Nevertheless, even the best theoretical concepts cannot always adequately solve practical problems in teams. The provision of computer-aided feedback on sprint team dynamics alone does not automatically lead to improved development performance. The development of improvements is dependent on the team's motivation and mentality, which, in turn, is dependent on several factors (e.g., performance-dependent bonus payments, acknowledgment, career prospects). Furthermore, the acceptance of new scientific methods or evidence by teams should not be taken for granted (e.g., by practitioners with decades of experience) [202].

> **The goal of the evaluation** is to assess the computer-aided feedback concept in two agile software projects carried out by practitioners. The evaluation strives to identify practical problems by the teams during the assessment allowing concrete actions to improve the usability and utility of the retrospective, predictive, and exploratory sprint feedback assets in ongoing projects.

The complimentary assessment of the usability and utility of the ProDynamics plugin in two industrial projects was motivated by previous studies, which solely covered the technological feasibility (see Section 5.5). The goal was to identify as yet unknown practical issues that may arise during the ASD process in the industry and that limit practical understanding of the socio-technical aspects of sprints and to make improvements accordingly. In this context, the evaluation was conducted in accordance with an action research process that involved two newly begun projects with distributed teams located in Germany and Spain. The teams had an open-minded attitude to new scientific methods, led to a better understanding of the socio-technical aspects in sprints without increasing the teams' workloads. The evaluation also covered different role-based perspectives and expectations related to the retrospective, predictive, and explorative sprint feedback support modules. The following sections present some background information on the projects and teams, the identified problems related to the usability and utility of ProDynamics, the resulting interventions for improvement, and the roadmap for the study.

6.1.1   Characteristics of the Projects, Teams, and Development Process

The use of the ProDynamics concept was evaluated in two agile software projects in a small IT company with about 40 full-time employees. The company currently has ten years of domain experience developing web applications (with a primary focus on custom-tailored software for human resources). The company was interested in participating in the evaluation study to try out on two software projects an innovative method for low-effort feedback support covering an extended summary of development performance combined with sociological changes in sprints. Since the ProDynamics plugin is an extension for Jira, the company's previous standards could remain unchanged (e.g., development processes and workflows).

The ProDynamics plugin facilitated the provision of supplementary feedback on the team behavior in sprints. No project constraints were identified in advance that might have had possible adverse effects on the evaluation. Table 6.1 summarizes the main aspects of the software projects with respect to processes, team structures, and technological infrastructure.

Table 6.1: Characteristics of the Two Evaluated Software Projects in Industry

Aspect	Description (short)
Projects	- newly begun projects (with one week offset) - estimated duration of one year - software releases in scheduled intervals
Processes	- agile development process (ScrumBut) - sprint duration between one and four weeks - sprint Retrospective at the end of a sprint - sprint Review at the end of a version-release - conditional sprint (QA) dependent on claims in pre-release
Teams	- newly assigned teams with three to six core members - all team member with multiple years of practical experience - distributed team members located in Germany and Spain[1] - defined roles (e.g., testing was only performed by the tester) - responsible IT head of teams was a senior developer - project manager functioned as product owner - developer had no direct contact with customers - project manager mediated between customer and developer
Infrastructure	- Jira was used for managing development task - Git was used as a central code repository - Jira was used to exchange task-related information - MS-Teams was used for ordinary scheduled team meetings

[1]Remark: All project members worked from home offices, due to the Covid-19 pandemic.

6.1.2 Action Research Cycle: Assessment, Intervention, and Reflection

The evaluation of the use of the ProDynamics concept in the two industrial software projects included qualitative assessments using an action research methodology. Action research is a pragmatic approach that involves diagnosing problems, defining an action intervention, and reflecting on the changes [11]. The methodology is driven by practical problems. It emphasizes participatory research and the iterative development of practical solutions [68, 183, 202]. However, the term "action research" is not clearly defined in the software engineering literature, as distinct from its basic definition in the social science literature [19]. In social science, action research is a unified process for social inquiry based on two stages: assessing the social situation in a practical context and deriving solutions to problems [18, 202].

During the last number of decades, this basic action research process has evolved in different disciplines, which has led to the more detailed canonical process shown in Figure 6.1 [72, 202]. Regarding the goals of action research, the five-stage process is still problem-focused and builds on participatory research. The collaboration between academic researchers and industry practitioners is an essential characteristic for a holistic situational assessment and a subsequent problem intervention [19, 72, 183].

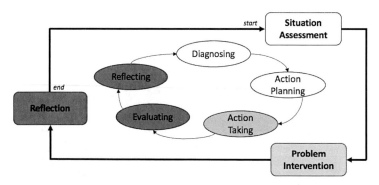

Figure 6.1: Action Research Cycle, cf. [19, 202]

The situation assessment included practitioners' perceptions and the problems they experienced with respect to the usability and utility of the individual sprint feedback modules in ProDynamics. The qualitative assessment was conducted at the end of a sprint (if possible, before or after a sprint retrospective to minimize its influence on the teams' development progress). The derived interventions were practically realized within the following one or two sprints, depending on the complexity of the changes required. The reflections were recorded and, after a time interval, made available for the teams in Jira (i.e., as part of a reassessment during the next appointed situation assessment meeting).

6.2   Action Research Process Applied in Practice

The action research process was carried out adaptively in accordance with the agile development process of the two industrial software projects. In this context, the socio-technical data capturing methods and the retrospective, predictive, and exploratory sprint feedback modules were individually assessed for their usability and utility by following the five-stage action research cycle (ARC).

The usability of the sprint feedback modules concerns the satisfactoriness of the outcomes (e.g., understandability, ease of learning, and readiness for use). Utility concerns the usefulness of the sprint feedback modules (e.g., whether the accessed module supported an understanding of changes in team communication in sprints). Figure 6.2 shows the report scheme that was followed to document the outcomes of the ARCs for each computer-aided sprint feedback module. The report scheme was adapted from the action research work conducted by Santos et al. [202]. Each of the five process stages covered additional subsections concerning related artifacts.

Assessment	Intervention	Reflection

Diagnosis
   i.   Context of Use
   ii.   Responses
   iii.   Quality Aspect
Planning
   iv.   Action Definition
	Action Taking
	   v. Refactoring
		Evaluation
		   vi. Reassessment
		Reflection
		   vii. Learning
		   viii. Remarks

Figure 6.2: Report Scheme Applied in the Action Research Cycle, cf. [202]

In this context, the situation assessment in the ARC covered two activities related to the individual sprint feedback modules: the diagnosis of problems and the planning of action to solve the problems. These activities were conducted through remote interviews at the end of sprints (preferably before or after sprint retrospectives). A set of assessment questions (AQ) was used to direct the interviews.

> *AQ 1: For what purposes were the modules used?*
>
> *AQ 2: Were the objectives of the sprint feedback modules understandable?*
>
> *AQ 3: Were the sprint feedback modules easy to use?*
>
> *AQ 4: Did the sprint feedback modules improve understanding?*
>
> *AQ 5: Did the sprint feedback modules support the project roles?*

The questions helped to identify the practitioners' perceived problems and experiences of each sprint feedback module in Jira. For example, the diagnosis stage concerned the context in which the feedback module was used.

The reported problems were classified according to the characteristics of the software quality model defined in ISO/IEC 25010:2011 [109]. The classification of problems was covered in the reports to ensure a shared understanding of the intervention to resolve quality issues.

> *"A quality in use model is composed of five characteristics (some of which are further subdivided into sub-characteristics) that relate to the outcome of interaction when a product is used in a particular context of use. This system model is applicable to the complete human-computer system, including both computer systems in use and software products in use"* [109]

The identified problems led to intervention actions, which were consecutively realized within the next one to two sprints, depending on the complexity of the refactoring. Reflections on the latest changes were discussed during the next sprint. Concerning the assessment, intervention, and reflection of the covered sprint feedback modules, it is essential to understand that the action research conducted does not preserve general repeatability of the outcomes, which is commonly the case in natural sciences research [176]. The repeatability of the action research outcomes is limited mainly because of the subjective nature of the qualitative approach (e.g., new experiences or varying perceptions of the practitioners).

In this work, the action research process represented a quantitative way of assessing the computer-aided feedback concept through practitioner feedback. This facilitated the identification of some previously unconsidered practical problems in the industrial ASD. One advantage of using a qualitative approach is that it enables the practitioner and the researcher to jointly explore a problem in all its complexity instead of having them make abstractions unilaterally [210]. The action interventions were defined based on theoretical and practical experiences from academia and industry [183].

6.3 Study Roadmap

The action research on the two industrial software projects followed a study roadmap for a systematic assessment and improvement of the socio-technical data capture and the retrospective, predictive, and exploratory sprint feedback modules with respect to their usability and utility. The roadmap describes a successive process of problem assessment, intervention, and reflection on the different sprint feedback modules in the ProDynamics plugin for Jira.

Practical problems can be revealed early in the newly begun projects, which mitigated follow-up risks in subsequent modules. For example, unsolved practical problems that hampered the socio-technical data capture would inevitably affect the retrospective feedback module's information foundation. This inaccurate retrospective information would, in turn, impact the accuracy of the predictive feedback support, and the exploratory feedback module would also be affected, since it

builds on the information provided by the retrospective and predictive sprint feedback modules. Figure 6.3 shows the study roadmap with the applied ARCs. The
ARCs were module-focused and followed a successive scheme that ran in parallel
to the progress of the two industrial projects for the first 18 weeks. The roadmap
structure was endorsed by the project management during a planning meeting concerning the expected use of the ProDynamics plugin (i.e., expected need for early
feedback on development dynamics in previous weeks and sprints during the first
half of each project due to the fact that teams were newly formed).

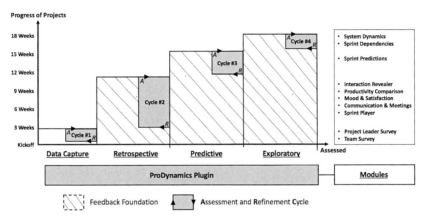

Figure 6.3: Study Roadmap for the Module-focused Action Research Cycle

The different time ranges for the ARCs were subjectively estimated due to the complexity of and diversity of the information in the ProDynamics feedback support
modules. The roadmap provides a general orientation. While the evaluation of the
project varied slightly over time, the iterative assessment structure did not vary.
The following subsections describe the ARC outcomes for one ProDynamics module to demonstrate how the action research was conducted.

6.3.1   Situation Assessment for the Productivity Module (ARC Excerpt)

The retrospective feedback covers a total of five modules. A situation assessment
for the retrospective productivity module, which is the first activity in the ARC,
is summarized in Table 6.2. The "Report ID 2.4" stands for the second ARC conducted (retrospective feedback). Number four refers to the productivity module
(for details about the module, see Section 5.2.2.4).

The situation assessment concerned the context in which the module was used,
while the problem-oriented practitioner responses related to the perceived quality
issues during its use. The problems were classified using the quality characteristics
of the ISO/IEC 25010:2011. The identification of the appropriate action for refactoring was a collaborative outcome that considered the practitioners' expectations
for improved retrospective feedback support concerning productivity in sprints.

Table 6.2: Situation Assessment Summary for the Productivity Module

Diagnosis	Description - Report ID 2.4
Context of Use	The module was used in the sprint retrospectives to examine the workload distribution and achieved velocity in relation to the regularly applied burndown charts.
Responses	*P2.4.1:* The visualization (radial chart) was reported to be insufficient for comparing the teams' development performance (see [139]). Moreover, a comparison of different sprint performance outcomes was not possible because the visualization only supported the selection of a single sprint at a time.
	P2.4.2: : The information covered in the visualization was reported to be incomplete with respect to the logging of tasks and work hours. Time logging (e.g., of stories) is a native function in Jira but was covered in the module because it was not used during the technical feasibility studies (see [128, 139]).
	P2.4.3: Feedback on the team workload and performance in sprints was reported to be generally valuable. Few members expressed concerns about possible comparisons by name. The privacy aspect is also relevant for the other ProDynamics modules that process member-related sprint information.
Quality Aspect ISO/IEC 25010	Usefulness, Context Coverage, Confidentiality, Maturity

Planning	Description
Action	*A2.4.1:* The visualization must support comparisons. Side-by-side bar charts would facilitate cognitive interpretation of productivity outcomes over time. They facilitated an evaluation of development performance and workload deviation within teams and across sprints. Mock-ups were used to create a shared understanding of the visualization format.
	A2.4.2: The task work hours of all team members must be visually presented and aggregated for each sprint. The bar chart visualization supports a line chart overlay that depicts work hours in sprints and teams, thus facilitating comparisons. Productivity changes and work hours could also be compared.
	A2.4.3: Team members must be able to adjust their privacy settings in their profiles based on their concerns regarding the visibility of sprint information to others within the team. Consequently, when a team member refuses to share personal sprint information, the data of all other members should also be anonymized to restrict comparability to project roles.

6.3.2 Problem Intervention for the Productivity Module (ARC Excerpt)

The problem interventions for the productivity module were practically implemented in the second instance of the ARC. The module was refactored in accordance with the collaboratively defined improvement actions from the situation assessment. The implementation lasted three weeks and was subsequently made available for the teams in Jira. A brief description of the implemented changes was provided for each problem addressed in the refactoring. ProDynamics plugin updates were announced in the projects through pop-up notifications in Jira. Moreover, all adaptions were individually communicated to the project manager in advance, as for the other modules. Table 6.3 summarizes the improvement actions corresponding to the actions specified in Table 6.2.

Table 6.3: Problem Intervention Summary for the Productivity Module

Action Taking	Description - Report ID 2.4
Refactoring	*R2.4.1* The refactoring included a redesign of the module's frontend visualization based on the improvement identified in *A2.4.1*. The radial charts, which depicted a team's productivity in a single sprint, were replaced with a sequential bar chart covering up to ten sprints on one time axis. The redesign enables direct comparisons of the team's development performance across sprints (cf. [139] and Section 5.2.2.4). The refactored main visualization includes a supplementary side view with a diverging stacked bar chart for a comparison of the development performances of a team (e.g., task-based workload variation among members).
	R2.4.2 Automated processing of a team's task-related work hours in sprints was integrated into the module's back-end routine. The productivity module visualizes the aggregated work hours of individual team members and those of the team for each sprint. In this context, the refactored bar chart visualization from *R2.4.1* was extended with a line chart overlay so that the team's work hours and development velocity across different sprints are cognitively comparable over time.
	R2.4.3 Privacy support was implemented for all modules. A team member can (de)activate the anonymous mode in Jira concerning the data processing for the retrospective, predictive, and exploratory sprint feedback modules. The mode was implemented to provide anonymity in the sprint feedback (e.g., mood and communication). To avoid exposing individual members, project roles are used in the visualizations instead of names.

6.3.3   Reflection for the Productivity Module (ARC Excerpt)

The last instance of the ARC covered a brief reflection on the project and a qualitative evaluation of the productivity module adapted to address the problems reported in the situation assessment. This involved a review of the module's initial problems and the changes applied to address the issues, followed by a reassessment. The primary goal was to determine whether the initial obstacles had been resolved or whether a further cycle was required to reinvestigate and address the issues in a different way.

In this context, the assessment questions (AQ 1-5) from the situational assessment were used again to reassess the adapted module. Finally, a reflection session summarized the central learnings from the ARC and general remarks from both research and industrial perspectives. The reassessment and reflection were carried out after the module adaptions were made available to the project teams in Jira at the end of the follow-up sprint. Table 6.4 summarizes the outcome of the reflection for the retrospective productivity module.

6.3.4   Assessment Conclusion for the Productivity Module

The adapted visualization and the extended information coverage provided computer-aided feedback support for discussions about development performance in sprint retrospectives, while also protecting the identity of individual members to address privacy concerns. The ARC resulted in practical improvements to address quality issues identified during an industrial study. However, the excerpt discussed in this chapter describes only one of several other ARCs conducted on the two industry projects. The ARCs addressed the computer-aided data monitoring module and the eight feedback-focused modules covered by the ProDynamics plugin. Future ARC assessments and interventions can benefit from ARC reports, as they represent a qualitative summary of the action research.

The overall goal of assessing the ProDynamics plugin in an industry setting was to assess and, if necessary, improve the usability and utility of the feedback modules in practice. The action research revealed multiple usability and utility flaws related to practical problems, which led to specific actions to improve the feedback support. Several further questions arose and were addressed based on individuals' observed development activities and sprint feedback. Does the ProDynamics plugin behave differently when applied in industrial processes than the behavior intended by design decisions resulting from academic environments? How do practitioners behave regarding their acceptance, actual use, and the perceived added value of the computer-aided sprint feedback in practice? Which information asset helped the teams most during sprints? Section 6.4 describes the quantitative and qualitative findings of the assessments of the use of ProDynamics assessment for the two industrial projects.

Table 6.4: Reflection Summary for the Productivity Module

Evaluation	Description - Report ID 2.4
(Re)assessment	The module was used in the context of sprint retrospectives, which included preparation for meetings and discussions about team performance in sprints.
	The adapted design of the module visualization clarified the objectives of the sprint feedback support. These changes addressed the comparability of development performances across different sprints and within the teams.
	The module was directly usable, but it was necessary to try it two to three times before the first sprint retrospective to learn all of its interactive features (e.g., hovering for extra information).
	The module supported the teams' understanding and awareness of performance outcomes through the visualized and processed development metrics, which had been less obvious before. This included a time-based comparison, which facilitated the identification of a significant performance change whereby the allow velocity in one sprint almost tripled in a subsequent sprint based on less task-related problems.
	The responses revealed no explicit role-related benefits except for the project manager (e.g., facilitated fast sprint feedback about past performances, thus the preparation of retrospectives). The module functioned as a shared information asset for the teams and promoted discussions and awareness of performance.

Reflection	Description
Learning	The use of the module was integrated into the teams' development routines as a performance-related feedback asset for sprint retrospectives. Compared to the previous observational study about technological feasibility, the ARC conducted in this study enhanced the maturity of the module, making it more useful and extended the information coverage and confidentiality for members with privacy concerns regarding their performance.
Remarks	The responses from the teams revealed that the manager used the productivity module most (e.g., for preparing retrospectives and screen sharing in meetings). In contrast, other team members passively accessed the feedback but found it useful to have it available when needed.

6.4   Quantitative and Qualitative Study Findings

This section describes the quantitative and qualitative results concerning ProDynamics sprint feedback support in practice. The findings are based on activity observations related to the use of the ProDynamics modules (e.g., time and frequency of accessed feedback by project role). In addition, the weekly captured socio-technical information revealed a set of team dynamics within the two agile software projects from the industry.

6.4.1   Observations Regarding the Socio-Technical Data Capture

Feedback that only concerns technical (productivity) sprint information cannot provide a holistic view of past or future team dynamics in agile software projects. Therefore, the computer-aided sprint feedback in this thesis builds on both the objectively tracked productivity metrics and the subjectively captured perceptions of individual team members regarding the sociological aspects of sprints (see Section 4.4.2). The motivation and willingness of teams to provide and receive feedback is important for continuous improvement. Therefore, general acceptance and an open mentality in teams are crucial for feedback on team dynamics in sprints.

The following observations are based on the socio-technical sprint data captured during the two industrial projects (e.g., provided data completeness by the teams over time). Figure 6.4 summarizes the participatory changes in the two projects and the resulting socio-technical data coverage for the first 18 and 17 weeks, respectively (until the first product releases). The timeline also depicts the duration of sprints for the two observed projects. For example, a sprint in Project 1 (P1) lasted one to two weeks, whereas the same sprint lasted two weeks in Project 2 (P2) (due to unexpected problems that arose in the previous sprint).

However, the quantitative summary shows that both projects involved a core staffing of three active team members, which increased to an overall high of active members midway into the first product release and decreased again near the end. The sociological surveys were sent at weekly intervals (automated routine in Jira) to all active team members, and the technical (productivity) data was tracked in Jira. An additional satisfaction survey was automatically sent to the project manager at the end of every sprint to assess the latest sprint performance (team and product-focused). Satisfaction reflections from customers were generally not accessible during the study.

The participation of all project members in the surveys was not mandatory. However, a high participation ratio increases the objectivity of the subjective responses, which in this case corresponded to the team dynamics feedback. On average, three of five active team members completed the sociological surveys in P1, and two of three team members completed the surveys in P2. Figure 6.4 depicts that the maximum data coverage was achieved for sprints 3 and 4 in both projects, whereby all active team members and the manager completed the surveys (100%).

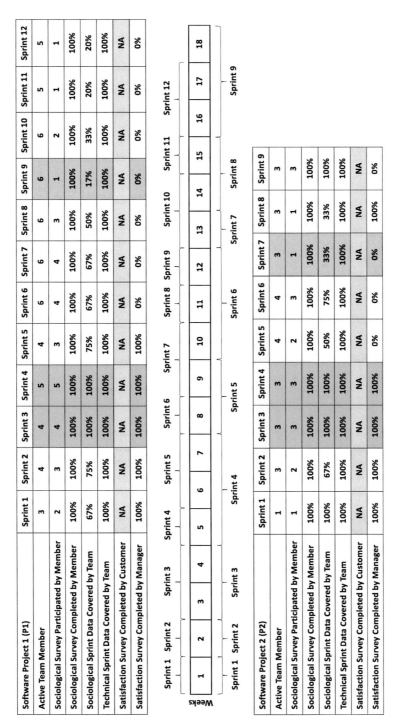

Figure 6.4: Socio-Technical Data Coverage in sprints with Participatory Changes

The automated information processing of the ProDynamics plugin builds on the motivation of team members to supply sociological data complementary to the passively tracked technical data in exchange for receiving extended (retrospective, predictive, and exploratory) sprint feedback support on the team dynamics in sprints. The more team members reflect on their status during a sprint, the easier it is to obtain an objective interpretation of socio-technical aspects of the team. A participation ratio of above 50% facilitates a comparison of the subjective perceptions of active members, thus facilitating the identification of significant deviations within teams and sprints (e.g., an overall mood change in the team or sole members).

Regarding the observations of the two industrial software projects covered in this chapter, two-thirds of the first product release schedule reached a socio-technical data coverage of at least 50% (on average >70%). In contrast, during the last third of the observation period, the participation ratio of the active team members and the manager stagnated significantly. Figure 6.4 shows the sprints with the lowest survey participation for both projects (marked orange). In sprint 9 of P1, only one active team member completed 100% of the socio-technical surveys, which means that the sociological data coverage reached only 17% of the maximum possible. In P2, the lowest data coverage (33%) was for sprint 7. The observational findings related to the socio-technical data capture in the industry are interesting because they are congruent with the observations from the technological feasibility study previously conducted on student software projects (see Chapter 4).

The previous study on the technological feasibility of the ProDynamics plugin conducted with academic teams had a similar participation outcome during the first two-thirds of the scheduled project duration (above 50%), while participation in the sociological surveys rapidly stagnated in the last third (between one and two active team members), which is similar to the observations in the industrial projects. A sample of $N = 100$ surveys in Jira resulted in a median completion time of 1 minute and 38 seconds. Lack of time due to increased workloads in the teams nearer the end of the first product release was repeatedly mentioned during the assessment interviews as the primary reason for the stagnated participation ratio. However, the general motivation within the teams to provide sociological information to assess the team dynamics in sprints as long as time constraints allowed them to do so was also repeatedly affirmed in the meetings. In contrast, the coverage of the technical data in Jira (productivity and progress measures) during all sprints of the industrial and academic software projects reached a constant level of 100%.

Observation Conclusion:
The case study of the two industrial software projects that use ASD shows that the concept for socio-technical data capture described in this thesis is applicable in practice. Nevertheless, the observations also revealed certain weaknesses concerning steady participation in the weekly sociological and sprint satisfaction surveys. Both strongly depended on teams and managers' workload and deadline concerns rather than motivational factors. As a result, a focus on more lightweight and practical methods that promote active capture of the sociological aspects of sprints is essential to reduce or even eliminate the need for responsive participation of members in periodic surveys.

6.4.2   Observations Regarding the Retrospective Modules in Practice

The conducted action research was motivated by RQ2 and RQ3, concerning the acknowledgment and perceived utility of computer-aided feedback on socio-technical changes in sprints. The goal was to obtain a better understanding of the inherent dynamics in agile development teams (see Section 1.2.2). The following observations resulted from 18 weeks of action research conducted on two industrial software projects. The findings are presented for each of the five retrospective sprint feedback modules introduced in Section 5.2.

6.4.2.1   Observations Regarding Module 1 - Retrospective Sprint Player

The module provided the two teams with retrospective sprint support for simulating and examining past state transitions in development tasks over time (e.g., stories, bugs, and sub-tasks). The retrospective player automatically detected the team-specific development workflows of the two project teams. It adopted those individually for the simulation viewer (i.e., transition states corresponding to a customized software development workflow on an agile scrum board). Figure 6.5 shows a similar development workflow of both teams based on the tracked task transitions in the two industrial projects.

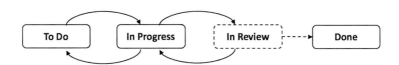

Figure 6.5: Extended Development Workflow in the Two Industrial Projects

A passive observation was made corresponding to the traceability of the development progress and the task transition states. The retrospective simulation of the development workflow with a time-lapse of 18 project weeks affirmed the general utility of an additional "In Review" state, which functioned as quality assurance before development tasks could change to "Done." The extended workflow in the two projects covered forward and backward transitions between the four progress states. However, the retrospective task transition simulations revealed that all development tasks in both projects were also changed to a never-revoked "Done" state when they passed the review. Thus, completed tasks resulted in no cases of reopened tickets, making the back-transition of the "Done" state dispensable due to the adapted workflow within both projects.

Table 6.5 summarizes the actual usage by the teams based on the quantitatively measured activities regarding the retrospective player module. The observation in the two industrial projects revealed limited use by the team members. Only 50% of the active team members in each project used the retrospective player during the 18 weeks of the observation (i.e., run simulations of task transition over time).

In both projects, the retrospective player was primarily used only during the first half of the total observation period and at the end of sprints. The static Scrum board in Jira did not natively support the reviewing of progress changes over time or the additions of ticket annotations for tracing problem comments related to a particular task (e.g., found during the "In Review" state and was changed back to "In Progress").

The subsequently conducted situation assessment concerning the context of use and recognized problems related to the usability or utility of the retrospective player revealed that the sprint retrospectives were always held remotely. Usually, a single member of each team (generally the project manager or IT head) remotely shared his or her screen with the other members during the retrospective meetings to review the previous sprints' performance achievements captured within Jira.

Table 6.5: Observed Usage of the Retrospective Player Module in Practice

	User in Team	Accessed Simulations
Project 1	3 of 6	9 in 12 sprints
Project 2	2 of 4	5 in 9 sprints

This shared access to the retrospective player through only one member explains the low number of observed simulation runs. The expectations were to determine the usage behaviors by the different team members. At the same time, both project teams' qualitative reflections on the first module's utility were positive, as the retrospective player facilitated a dynamic review of task transitions states and progress changes during a previous sprint.

The findings regarding the qualitatively assessed utility and quantitatively observed usage by the teams of the problem annotation support in tickets were somewhat contradictory. The function was described during the interviews as a useful extension in Jira and was even improved based on the team feedback from the action research. However, it was only used once by each team during the entire observation period.

The summarizing reflections of both project teams at the end of the study revealed the reasons for the lack of use of the first module, particularly during the second half of the study. The retrospective player was perceived as a complementary sprint feedback module that was needed only in situations where additional simulation visualization was required to retrospectively review the progress of development tasks to discuss problems in a sprint. During the development process, the general lack of time during meetings for a detailed review of task-related progress (changes) in combination with the project managers' perception that "the teams performed smoothly, so there was no expected benefit for accessing extra feedback in the sprints" meant that the involvement of the retrospective player in most of the sprint retrospectives was dispensable.

6.4.2.2   Observations Regarding Module 2 - Team Communication

The second retrospective module supported both teams every week by providing additional feedback on their communication and meeting behavior during sprints. The team communication module was assessed and improved based on the teams' responses regarding the usability and utility of the retrospective module during the ARC. The module provided network graphs for characterizing and depicting each team's organizational changes over time (e.g., the centrality of individual members, the appearance of mavericks, and the interconnection between members).

Additional line charts visually summarized relevant communication and meeting metrics on a time axis (e.g., communication distance in teams, use of communication media, and the frequency and duration of meetings during sprints). This allowed to review and compare changes in communication over a maximum of ten sprints (see concept in Section 5.2.2.2). The team communication feedback was updated at weekly intervals and was directly accessible by the teams in Jira. Table 6.6 shows the observed usage of the communication module in the two projects.

Table 6.6: Observed Usage of the Communication Module

Project	User in Team	Accessed Communication Feedback
1	2 of 6	18 in 12 sprints
2	2 of 4	10 in 9 sprints

The team communication module was regularly accessed by both teams during the study, mainly at the end of a sprint and, in P1, after a highly productive week (in terms of development velocity). However, only two members accessed the team communication feedback individually, one of whom was the responsible project manager. The two members who frequently accessed the communication feedback during the sprints were also the most central persons within their teams (in terms of centrality). These observations were based on the weekly visualization of communication structures in the form of network graphs. The time points and frequency of the teams' use of the module confirmed their reports that they used it for routine communication as an integrated activity in the sprint retrospectives.

The low number of users was observed again due to the decentralized sprint retrospectives shown on a remotely shared screen by one team member for a joint discussion on the sprint-related performance outcomes recorded in Jira. The geographical distribution of both project teams led to increased social interactions between the members in sprints. This included a significantly higher demand for remote communication channels than observed in the centralized student software projects in the technological feasibility study (see Section 5.5). In this context, the qualitative team reflections indicated several instances when the retrospective module improved understanding of the previous weeks' communication and meeting diversity based on the sociological information gathered. Some of the reported benefits are described in Table 6.7.

Table 6.7: Reported Utility of the Communication and Meeting Module

Module Aspect	Project	Description
Organizational	1	A new developer joined the project after four weeks. The following two weeks showed barely any communication with the project manager. The manager used the communication module to determine whether the new member was integrated into the team or showed loner characteristics. It turned out that he optimally integrated himself into the team structure.
	1, 2	Interim reviews of the team structures revealed that the teams' communication structures involved people who were not active team members of the respective projects. Additional feedback revealed that some company experts were occasionally contacted for help and were thus captured in the networking graphs.
Communication	1	The perceived communication intensity increased significantly during sprint 5 compared to the steady course over the previous weeks. The project manager knew that the productivity module included nearly twice the number of development tasks as in the previous weeks but was not aware that the communication in the team also intensified.
	2	Sprint 4 contained significantly more planned development tasks compared to the previous sprints. The communication behavior and development productivity were at an average level in the first week of the sprint. The module revealed that the team structure changed to an unusual solitary structure in the second week, after which the productivity of all developers almost doubled productivity in the next week of the sprint.

In contrast to the technological feasibility study on student software projects, this study did not explicitly assess the utility of the module for statistical inferences (e.g., measurable impacts on development performances or organizational changes based on the socio-technical feedback) [136, 139]. Reasons for this included the high variability in active and inactive team members, the variable duration of sprints, and the erratic satisfaction feedback from the project manager and customer. Nevertheless, the communication feedback in the retrospective module supported the characterization of both teams' communication network based on the observations related to the lowest, medium, and highest communication behavior (in terms of communication intensity and connectivity). Figure 6.6 and 6.7 show the observed team communication extremes corresponding to the average team communication among all sprints.

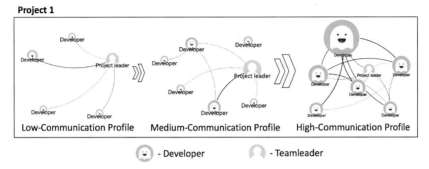

Figure 6.6: Observed Team Communication Network in Project 1

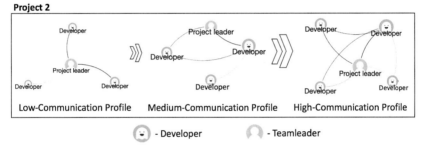

Figure 6.7: Observed Team Communication Network in Project 2

The interviews with the teams during the ARC revealed a general utility problem that also applied to all ProDynamics sprint feedback modules concerning the availability of active intervention recommendations based on experiences from previous sprints. For example, the Flow Distance showed an overall high value in this case study, which implied that different communication channels were frequently used in the teams. Due to the Covid-19 pandemic, the teams communicated and exchanged information exclusively via digital media channels, leading to a high flow distance. In contrast, for more centralized teams, different information channels (e.g., face to face) might be a more effective communication format.

A definitive meaning of the high flow distance in the teams cannot be determined because, according to their project managers, both teams performed smoothly without significant problems. Consequently, an active intervention recommendation also depends on the external team and environmental factors, thus it is not always trivial to automatize. Moreover, it does not suffice to provide technical guidance or intervention recommendations without considering additional feedback on the positive or negative perceptions of the communication; instead action recommendations should be based on the situation.

The retrospective communication module was recognized by both teams as the most useful one, and this was also reflected by the observed usage. The sociological insights promoted awareness among team members of the team dynamics (e.g., with respect to dependencies between team structures, communication behaviors, and sprint productivity). Moreover, the situation-dependent utility of this module was underlined by individual statements of members (e.g., "helpful for new teams or in ongoing projects with new on-boarding members" and "due to the small size of the teams, no benefiting findings were expected. The module is probably more useful for larger team structures."

6.4.2.3   Observations Regarding Module 3 - Mood and Satisfaction

An overall balance in the mood of the development teams, as well the satisfaction of customers and managers regarding the teams' performance during sprints, are crucial for team and project success [83, 96, 180]. Situational events that trigger mood changes (either positively or negatively) are relevant to the teams' awareness and understanding. Furthermore, every team reacts differently and thus requires an independent interpretation of its own socio-technical effects (e.g., two teams can perform similarly well in their projects but with an offset between their moods).

The third retrospective module supported both teams during this study through weekly summaries of actual mood changes (both positive and negative) through-out the projects, based on the anonymously shared perceptions of the team members. Moreover, the objective was to provide the teams at the end of each sprint with systematically obtained and thus comparable feedback on the project manager's satisfaction with the teams' performance and product outcome (customer feedback was not available for this study). The mood and satisfaction module was the second most frequently accessed feedback asset during the 18 weeks of study (the communication and meeting module was the most frequently used). Table 6.8 presents a quantitative summary of the observed usage of the module.

Table 6.8: Observed Usage of the Mood and Satisfaction Module

Project	User in Team	Accessed Mood Feedback
1	4 of 6	15 in 12 sprints
2	2 of 4	10 in 9 sprints

Both teams regularly accessed the mood module at the end of a sprint and in P1, as well after an irregular event (e.g., change requests). Two members with frequent access to the mood insights were the manager and IT head of each project. In P1, two more members accessed the feedback to compare their moods with the team average. The generally small number of users and low access rate were a result of the mood changes being reviewed remotely during the retrospective through one member's shared screen. The observations revealed a comparable average level of the positive and negative moods for both teams.

Figure 6.8 illustrates the mood for weeks with maximal imbalance compared to the observed average for each project over time. The representation of mood artifacts through radial charts helped the teams to identify emotional changes, such as the reception of an exciting announcement. Since other concepts typically only include positive or negative statements, it is difficult to make detailed conclusions on the project. The weighting of positive and negative mood attributes (emotions) resulted from each team member's weekly perceptions (see Section 5.2.2.3).

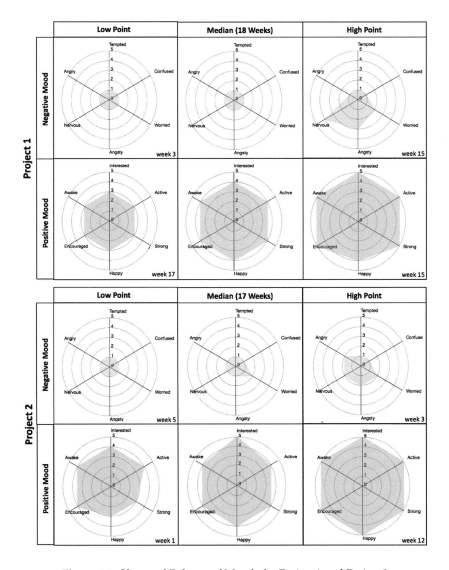

Figure 6.8: Observed Balance of Moods for Project 1 and Project 2

A small deviation between the perceived low and high points of emotions to the median implies a balanced mood. The low and high points of the positive mood during P1 and P2 did not differ significantly from each median, thereby revealing an overall steady positive mood balance. The team of the first project (P1) perceived the greatest positive emotions during the 15th week, while the other team (P2) experienced its emotional peak during the 12th week. In both projects, the low points of the negative mood did not significantly differ from each median either. In P1, the team perceived the lowest negative emotions in the third week, whereas In P2, the team perceived a negative emotion low point in the fifth week and a high point in the third week. Thus, P2 revealed an overall steady negative mood level.

In contrast, a significant change in the team's perceived negative mood was observed in P1 in the 15th week. The team exhibited a substantial increase in nervousness and angst compared to the negative emotion perceptions from the prior weeks. However, the team's increased negative mood was unsurprising, because it was the week before the last sprint of the first product release for P1, and the latest test coverage did not fully cover all functions at that time. The team was concerned about keeping up with the scheduled delivery date. The satisfaction ratings for the performance of both teams for each sprint were constant without significant deviations over time, which affirmed the project managers' reflections of an overall smooth development. Nevertheless, the emotions of individual members or teams can change significantly based on particular events in P1 in the 15th week. To understanding the emotional changes, it was essential to observe the various mood characteristics, even when the performance outcomes of the teams were steady. Otherwise, changes in mood or event-triggered impacts cannot be systematically determined after the projects.

The module was assessed for the two projects and improved during the action research corresponding to the practitioners' perceived usability of the module and of the available mood and satisfaction feedback during the sprints. The team reflections uncovered a conceptual problem regarding the systematic capture and provision of feedback for the negative moods in teams within the ProDynamics plugin. During an interview, the manager of P1 mentioned that two team members had a bilateral (social) conflict, which escalated more than once during the sprint retrospectives and even required some management intervention. The weekly feedback on the positive and negative moods did not indicate signs of social conflicts within the P1 team. The negative emotions mentioned were not covered because these two members never completed the weekly sociological surveys, and both of them displayed a functional and positive working atmosphere with the rest of the team. Therefore, no negative perceptions were reported by the other members as well. Moreover, the survey only captures personally perceived emotions. Consequently, these two team members presented a blind spot based on the missing sociological data, thereby omitting feedback that would expose the dysfunctional situation in the team. The issue threatens the validity of the usefulness of the module and is discussed in the related section.

6.4.2.4   Observations Regarding Module 4 - Productivity Comparisons

The objective of the productivity comparison module is to support the teams through fast feedback about the development velocity achieved by teams and individual members. At the same time, it enables productivity comparisons over time to determine performance courses and trends. The module provides retrospective sprint feedback about the teams' development diversity over time. It visualizes the development performance based on estimated, completed, and incomplete tasks and story points (i.e., for the team and its members), thus allowing comparable velocities across sprints. Regardless of the type of performance metric (e.g., velocity, or work-hours per task), organizations need to consider the average from the past as a possible estimation target during the sprint planning as one of the principles of the Agile Manifesto is to promote sustainable development through a steady pace that clients, developers, and users can keep for an unlimited period [24]. The module was assessed and improved in an ARC concerning the feedback usability and utility (e.g., extension to regular sprint reports, burndown charts as commonly used on agile software projects). The activity observations related to the quantitative usage of the module in the two projects are summarized in Table 6.9.

Table 6.9: Observed Usage of the Productivity Comparison Module

Project	User in Team	Accessed Productivity Feedback
1	2 of 6	12 in 12 sprints
2	2 of 4	7 in 9 sprints

The activity observations regarding the module revealed that the project manager primarily accessed the productivity feedback in Project 1, and Project 2 disclosed equivalent usage by IT head and project manager. In both projects, the module was primarily accessed during the sprint retrospectives or in advance by the project manager, whereas the low number of total access depends on the teams' way of working (i.e., shared remote desktop in the team meeting). The productivity comparison module revealed a significant imbalance of the velocity-based development performance throughout each project. Figure 6.9 summarizes the relative performance courses and trends over time.

Even though the teams were aware of the resulting performance highs or lows at the end of the sprints, the summary of the comparable (across) sprint visualization reminded some members of how significant the performance gaps are. However, the project-related trendlines functionally characterize the performance directions, which pointed out a seemingly ideal sprint performance increase towards the due date of the first product release (P1=91% and P2=98%). A reason for this is acquiring additional team members (HR-based capacities) to get the performance back on track, which in both projects resulted in significant fluctuations regarding the number of regular active team members and sprint-wise associated team support. This resulted in a high diversity of the workload balance in both teams, far from a steady pace (see example in Section 5.2.2.4).

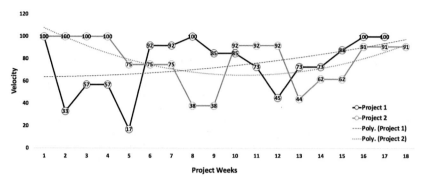

Figure 6.9: Velocity-Based Performance Variation and Trends in the Projects

However, not the number of additional members caused a positive performance change, but the number of working hours these additional members got approved to spend in the project. The development observations contrast the repeatedly reported smooth project course of both projects, whereas the qualitative response by the management implies the high subjective nature for interpreting team performances.

6.4.2.5   Observations Regarding Module 5 - Interaction Revealer

The task-related interaction revealer is the last in the chain of retrospective feedback modules. The module supports, likewise to the previous team communication module described in Section 5.2.2.2, the visualization of team interaction structures in sprints. However, the communication network depends on the team contribution of sociological data needed to analyze and visualize the communication networks. In contrast to the team communication module, the interaction revealer can monitor team interaction, primarily through task-related activities associated with development tasks. In Jira, specifically, these are file attachments, commenting, task (re)assignments, and forwarding [149]. During the action research, both teams disclosed a strong culture for communicating task-related requirements, instructions, and advice through the associated comments fields (e.g., in story tickets). In addition, task forwarding and assigning to others were standard practices used by all members throughout the sprints as well. Table 6.10 lists the activity observations related to the quantitative usage of the module in the two projects.

Table 6.10: Observed Usage of the Task Interaction Module

Project	User in Team	Accessed Task Interaction Feedback
1	2 of 6	9 in 12 sprints
2	2 of 4	7 in 9 sprints

Remarkably, the usage observations of the interaction revealer module revealed that both teams utilized the additional feedback similar to the previous feedback modules (i.e., the context of use was mainly in sprint retrospective and within the first two-third of the first release schedule).

The passive capture of all task-related activities in both projects supported the detailed characterization of interaction networks of both teams concerning their development activities over time. It enabled the teams to examine significant structural changes in their interaction behavior visually corresponding to the usual interactions throughout the projects. Figure 6.10 and 6.11 show the objectively captured task-related interaction networks of both teams by their structural extremes compared to the averages.

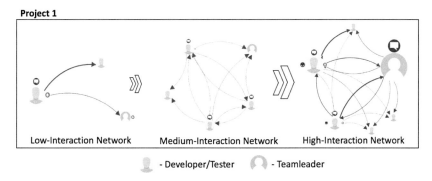

Figure 6.10: Observed Interaction Networks in Project 1

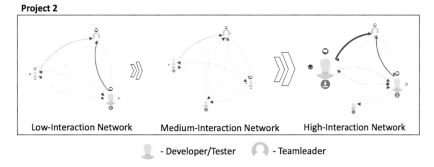

Figure 6.11: Observed Interaction Networks in Project 2

In contrast to the team communication networks in Section 5.2.2.2 does the interaction revealer module considers objectively tracked activities related to development tasks. Consequently, using this method, other involved members (e.g., based on pair-programming) cannot be determined. Nevertheless, its strength is being independent from the contribution of teams for sociological data. Therefore, the combination of subjective communication data and objective interaction tracking provides a complementary characterization of team structures.

Comparing the interaction extremes of both projects clarifies that the task-related lowest and highest interaction network in project 1 has a significantly more substantial structural deviation from the average than the other project. Project 2, on the other hand, shows a much more even interaction network, in which the extrema difference is primarily the intensity of interaction.

The module was assessed for the two projects and improved during the action research corresponding to the practitioners' perceived usability and utility of the interaction module during the sprint retrospectives. The team reflections uncovered only minor usability issues regarding the initial performance problems when loading the visualized interaction networks (e.g., delay when opening the module and switching between sprints to examine another visualized interaction network). The problems were refactored and reassessed within the next sprint, thus satisfying both teams' usability and utility expectations afterward.

6.4.3   Observations Regarding the Predictive Sprint Support

The action research with two agile development teams from the industry disclosed an active utilization of the descriptive (retrospective) sprint feedback assets during the sprint meetings throughout the projects. The five underlying feedback modules supported them to understand previous behavior patterns in socio-technical data (e.g., team communication, moods, and task-related interactions, productivity). At the same time, the supplementarily provided visualizations supported the cognitive understanding and awareness for arisen behavioral changes and significant trends regarding these socio-technical aspects. Thus, the retrospective modules satisfied the central goals of this work towards a practical understanding in teams for behavioral dynamics in ongoing software projects (see Section 4.3).

The predictive sprint feedback asset strives to complement the retrospective feedback information by proposing the predicted sprint performance for the following weeks considering each team's unique development behaviors through the project. Team behavior can change throughout projects, and so do socio-technical dependencies in sprints. The ensemble of five ML models takes these changes into account covered through different regression algorithms for analyzing the functional relationships in the socio-technical data (see Section 5.3.3). The predictive feedback asset is applicable in student projects and industry, as shown in the following Figures 6.12 and 6.13.

Figure 6.12 shows that SVR achieved the overall best data interpretation throughout the project 1 compared to the other models. However, due to some weeks, the dynamics selection even lowered the average RMSE by 1.1%. Figure 6.13 reveals that KNN achieved the overall best data interpretation throughout the project 2 compared to the other models. However, due to some weeks, the dynamics selection even increased the average RMSE by 0.7% (by mean, a static selection of the KNN predicts better in average regarding the socio-technical data of project 2).

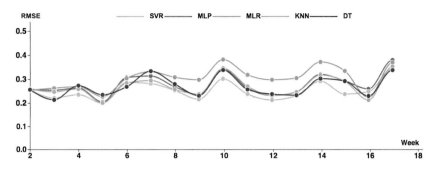

Figure 6.12: Predictive Sprint Feedback Validation of Project 1

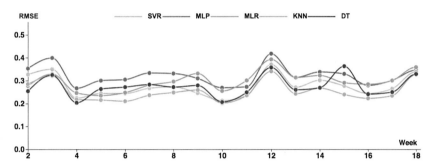

Figure 6.13: Predictive Sprint Feedback Validation of Project 2

The detailed validation results show the weekly prediction RMSE for the automated model selection routine. Remarkably, the dynamic selection compared to
the static use of the ML models show in almost all cases an improvement summarized in Table 6.11. The validation results indicate that the predictive feedback is
applicable for socio-technical data captured in industrial software projects, while
the average RMSE was equally low as for student software projects (see Section
5.3.3).

Table 6.11: Validation of Dynamic Model Selection versus Static

Project	MLR	KNN	MLP	DT	SVR
1	+4.7%	+1.5%	+2.3%	+1.6%	+1.1%
2	+4.5%	-0.7%	+7.4%	+3.5%	+2.5%

While a first comparative study about the technological feasibility of the concept
even resulted in statistically significant performance increases in student software
projects, this action research with two industrial projects aims to assess (and possibly improve) the perceived usability and utility from practitioners. Table 6.12
summarized the accesses of the predictive sprint feedback asset in both projects,
which matches the usage outcome reflected during the assessment interviews in
the conducted ARC.

Table 6.12: Observed Usage of the Predictive Module

Project	User in Team	Accessed Predictive Feedback
1	1 of 6	2 in 12 sprints
2	0 of 4	0 in 9 sprints

The predictive feedback was only considered twice by the project manager of project 1. The prediction support was included during the sprint planning. However, the interviews revealed that the team did not see the added value of forecasting for their performances because it lacks concrete instruction of what and how changes must be addressed to achieve improvements. As a result of the ARC intervention, descriptive implications for specific courses and deviations were defined and refactored to enhance the utility. However, while the improvement was acknowledged during the reassessment, the missing incentives remained in the teams that the asset gained no further attention. For example, how to improve anger in teams? The problem is nontrivial to answer and understandable to seek in a recommending system, but that goes along with various other issues (e.g., requires a substantial extension through recursive learning loops in advance of AI).

6.4.3.1   Observations Regarding the Explorative Sprint Support

The sprint dependency module builds on exploratory data analyses to visualize, in combination with network graphs, the socio-technical dependencies in and across sprints without limitations to solely linearity dependency analyses (see Section 5.4.1). In a previously conducted study about the technological feasibility of the sprint dependency module in six student software projects, the module was capable of disclosing linear and non-linear behavioral patterns within four sprints [134]. The applied action research in this work aimed to assess the utilization and utility of the sprint dependency module in two industrial projects, as well as the scalability of characterizing socio-technical data from student to industrial software projects. Table 6.13 summarizes the utilization of the module in both industrial projects, a prerequisite for the ARC team reflections about its utility.

Table 6.13: Observed Usage of the Sprint Dependency Module

Project	User in Team	Accessed Sprint Dependency Feedback
1	1 of 6	2 in 12 sprints
2	0 of 4	0 in 9 sprints

The technological feasibility of characterizing the socio-technical dependencies in data from the industry is possible, as shown in Figure 6.14. Several linear and non-linear sprint dependencies were analyzed and visually abstracted as network graphs accessible to the two teams. For example, Figure 6.14 reveals the significant influences on the team leader satisfaction throughout all sprints.

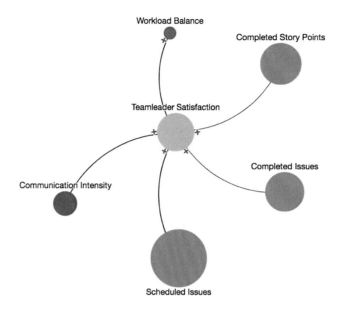

Figure 6.14: Significant Dependencies Throughout Nine Sprint in Project 1

The number of which the sprint dynamics module was accessed revealed no considerable application in the projects (similar to the sprint prediction module). The utilization of the sprint dynamics module, shown in Table 6.14, was no different. During the ARC for the exploratory feedback assets with only the project manager attending (the one who also accessed both modules at least twice), it turned out that the development process barely spare time to explore sprints dynamics and dependencies in such detail. Moreover, the budgets are tight, and the developer also has no extra time to use these two and the sprint prediction module actively. For an additional team member only responsible for QA tasks, this might result in a different utilization.

Table 6.14: Observed Usage of the System Dynamics Module

Project	User in Team	Accessed System Dynamics Feedback
1	1 of 6	2 in 12 sprints
2	0 of 4	0 in 9 sprints

Consequently, while the interviews about the predictive and exploratory sprint feedback assets revealed a lack of time and resources to apply detailed analyses during the development process, the question concerning of the utilization in these two industrial development teams remains unanswered for both feedback assets.

6.5   Threats to Validity

The results of this evaluation are only valid to a limited extent and may not be generalized. According to Wohlin et al. [241] this case study is subject to specific limitation of validity, which are divided into conclusion, internal, construct and external validity.

6.5.1   Conclusion Validity

The case study observations and findings during the evaluation of the computer-aided sprint feedback concept in industrial software projects are limited in generalization. The approach investigated the utilization and utility of the sprint feedback by only two agile development teams, which represents a small sample, insufficient to conclude a general transferable team behavior for other software projects.

Moreover, the two teams utilized not all sprint feedback assets during the observation, which requires further observations concerning the usage and utility of the predictive and explorative sprint feedback before a general conclusion is possible. While prior observations in student software projects revealed promising findings, it is about to determine the general utility of the computer-aided sprint feedback by practitioners for those sprint feedback assets that lacked utilization in the current evaluation.

6.5.2   Internal Validity

The internal validity of the evaluation is influenced by the fact that no control teams from other companies were involved during the study. Consequently, it biased the interpretability of the observed team behavior from a neutral perspective since no other references are known.

Since the evaluation was conducted in globally distributed teams, the Covid-19 pandemic may have affected the observation findings over time due to changing domestic conditions and restrictions (also concerning the participant feelings and behavior, which are a central focus of this work). Such an unrelated event can influence teams' behavior concerning the development activities and the participation in the study.

Another natural influence on the study observations due to the long period can be the maturation, by mean personal changes or willingness of the participant based on weekly sociological surveys and repeated research assessments exceeding their regular project tasks. Moreover, due to the high team fluctuation, individual members joined the study late or dropped out without further notice, which means that the observed team behavior may be distorted by the few members that permanently had a spot in the project.

6.5.3 Construct Validity

Qualitative assessment during the action research is solely based on the individual team members' perceptions and experiences. While some quantitative measures are involved in affirming the subjective statements (e.g., perceived smooth development performance and measurable progress metrics), the teams' utility feedback still lacks objectivity.

Another threat to construction validity is the linguistic and cultural bias during the meetings. Questions and other information in written form within ProDynamics supports multiple languages to prevent false interpretations (e.g., the team survey). While the team members have different international origins, soft language barriers may influence participants' understanding of questions during verbal communications.

Moreover, cultural aspects regarding the mentality to contribute or share the personal mood or perceptions in the team meeting were not considered for conceptualizing the computer-aided sprint feedback concept (e.g., the socio-technical data capture). That includes the different legal-law aspects across the European borders regarding the processing and saving of personal information in addition to domestic privacy restrictions.

6.5.4 External Validity

The external validity of the evaluation is always affected when the conditions in the study do not match the natural conditions for which conclusions are to be drawn from the observations. This includes the participants' commitment and personality, the different environmental conditions in the software projects, and influences during the development process.

Besides, Hawthorne effects were particularly possible throughout the observations in the projects. The individual team members might change feedback utilization behavior simply because they knew the study context. Nevertheless, the two teams impersonated the target group of agile software development teams to evaluate the computer-aided sprint feedback (ProDynamics), as they have different prior expertise, project roles, and perception, thus reflecting different future team members using ProDynamics.

6.6 Discussion and Conclusion

Data on team behavior should be simplified to track, analyze, and interpret as sprint influences are essential to understand [134]. The evaluation with two agile development teams from the industry showed the applicability of the (1) Socio-

Technical Data Capture Concept, the (2) Descriptive, (3) Predictive, and (4) Exploratory Sprint Feedback Assets in practice with limitations. Both teams showed an alternating number of active members per sprint P1 (3-6) and P2 (1-4) who contributed socio-technical data during the sprints (concerning the performance satisfaction feedback by the manager and sociological data by the team).

Although the data contribution in both teams stagnated near the end of the first product release, the observations showed that the subjective measures (satisfaction-related data by the project manager and communication, mood in the teams) and objective measures (tracked development progress and task-related interaction in Jira) could be both successfully collected (automated opinion polls) during the development process through ProDynamics. A similar participation pattern was observed in prior studies with student software projects. Nevertheless, further studies are needed to investigate whether this stagnating pattern near the end of the project or the first release also arises in other projects. The current sample of two industrial teams is too small for a reliable conclusion.

The evaluation aimed to determine whether the computer-aided sprint feedback concept scales up from the early approaches in student software projects to the unfettered development environment of industrial project. ProDynamics automatically processed the socio-technical sprint data in weekly intervals, accessible for the two development teams in the form of descriptive, predictive, and exploratory sprint feedback assets. The applied action methodology allowed ad-hoc improvements of the sprint feedback assets to be collaboratively defined and developed through a fast feedback cycle concerning the added value, with the prerequisite for the active usage of modules to know their actual utility.

Consequently, the action research with industrial software projects emerged several practical improvements through qualitative feedback from both development teams about the mainly utilized descriptive sprint feedback. Thereby, structured interviews assessed practical criteria to determine the context of use and utility (including noted problems). Moreover, the collaboration with the industry was essential to identify practical problems not disclosed during the prior studies with student software projects.

The jointly defined practical improvements of the five descriptive sprint feedback modules were positively affirmed through the regular utilization of the complementary feedback as a discussion base during the sprint retrospective with all members. The retrospective information visualization and characterization of team communication behaviors, mood changes, task-related interactions, and performance alternations in sprints provided both teams more awareness, thus a better cognitive understanding of the team behavior courses throughout the projects. Both teams' members barely accessed the descriptive sprint feedback individually rather than through remote desktop sharing during the retrospective event, which helped discuss performance influences. This usage behavior was unforeseen during the construction of the study and definition of assessment measures but is plausible because both teams' members are geographically separated and thus always meet up for these sprint retrospectives remotely.

In contrast, the predictive and exploratory sprint feedback found no considerable utilization in the two teams throughout the courses of projects. The missing context of use and incentive for the sprint prediction module, the sprint dependency module, and the sprint dynamics module reported during the action research assessment interviews by the project manager does not allow an objective conclusion about the utilization or utility for these three ProDynamics modules. Moreover, the lack of time for "exploring" sprint dependencies parallel to the projects' workloads hampered modules' utilization.

A bilateral conversation with the one project manager who used all three modules twice affirmed the missing incentive for actively involving the module in the development process because of no active recommendation of changing future courses. Nevertheless, the prediction model's data-driven validation results confirmed being reliably applicable in student and industrial software projects, without accuracy difference due to different team behavior in sprints. That does not allow concluding the usefulness in industrial software projects.Regarding the descriptive sprint feedback support, the teams perceived it helpful during the team formation in the newly started project. However, when time passes and the team perceives well without specific events, the interest and effort for contributing sociological data are reduced. Since the evaluation involved only two industrial projects from the same company, it is advisable to conduct further study studies with different development teams for a more objective, thus comparable assessment. In the context of action research, this could, at the same time, improve the degree of maturity of ProDynamics towards a more generalizable use.

Chapter 7

Conclusion

With the growing complexity of many software projects, software development is becoming increasingly dependent on teamwork. Consequently, the development process also requires consideration of the human factors in team contexts. Moreover, to achieve optimal development performance throughout a project, it is vital to understand the individual behaviors of teams over time, which allows for the identification of dysfunctional communication and interaction patterns and changing moods at an early stage. Information technologies can assist in the understanding of team behavior during the development process. Modern data methods enable the systematic capture of relevant socio-technical data and the analytical interpretation of behavioral changes.

Given the context of this thesis, the first research question addresses the practical relevance of supplementary team feedback on the socio-technical aspects of agile software projects. The question was investigated as part of a design science research methodology applied to prototypically realize computer-aided sprint feedback on team behavior in agile software projects with the help of integrated capturing methods and the direct analytical processing of socio-technical data in Jira. Moreover, the methodology included the identification of relevant information needs in agile development environments, taking into consideration practitioner experiences and related work from software engineering, organizational and social psychology, and data science.

In the last decade, the literature on the identification of related socio-technical metrics has revealed an increasing research focus on human factors in software engineering. However, this positive trend presents a limited indication of the practical relevance of socio-technical feedback in software projects. A survey study involving 90 researchers and practitioners reveals the most frequently encountered team issues in sprints. Among other aspects, 63% of the participants regarded a lack of "communication," "quality thinking," "quality skills," "misperceptions," and "withholding information" as team-related problems that affect the success of the sprints. Furthermore, 65% believe that these problems would have been avoidable through adequate team feedback. 86% of additional 69 responses disclosed that understanding team behavior in past sprints is crucial for improvements in planning for the next sprint.

The concepts for computer-aided sprint feedback support that emerged during the design phase were prototypically implemented into the ProDynamics plugin for Jira. An initial technological feasibility study with 15 relevant student software projects found performance improvements in seven teams (based on the perceptions of customers and team managers) that actively accessed the retrospective and predictive sprint feedback support in ProDynamics. These seven teams also exhibited a balanced velocity distribution with fewer estimation errors ($\pm 9\%$) compared to the other teams ($\pm 19\%$). Moreover, 77% of all the members of the seven teams recognized positive effects in follow-up development performance due to the additional feedback in ProDynamics.

Answer to Research Question 1

The applied design science research reveals the practical relevance of and demand for supplementary feedback on the socio-technical aspects of agile software projects with some constraints. Agile development teams that actively strive to improve their performance can better understand coherent team behaviors through computer-aided sprint feedback. However, the practical relevance of socio-technical feedback relies on the individual culture and commitment of the teams. Therefore, the findings should not be overgeneralized for other teams.

While the practical relevance of computer-aided feedback is a vital aspect of this investigation, it is equally important to examine practical usage in ongoing projects. In this context, the second research question of this thesis focused on how agile teams utilize socio-technical feedback during the software development process. Action research enabled quantitative and qualitative observations and cyclical team reflections regarding the use of the retrospective, predictive, and exploratory feedback modules of ProDynamics in two new industrial software projects. The iterative action research process promoted the early recognition and improvement of practical issues reported by the two agile teams, which did not arise during the design science research involving student projects.

The use of ProDynamics feedback was quantitatively observed for frequency, access times, and weekly participation in the sociological surveys. Both teams adopted the complementary feedback assets in sprint retrospectives as foundations in the discussion of performance deviations while considering sociological factors (primarily communication and mood changes corresponding to the level of productivity achieved during a sprint). Management regularly accessed the retrospective feedback module in sprint retrospectives or as part of the preparations. The Management accessed the predictive and exploratory sprint feedback only once or twice each for curiosity reason (e.g., to predict the course of mood following a poor sprint performance). The teams mainly considered the computer-aided feedback to stabilize their developmental performance, particularly during the first two-thirds of the schedule since the beginning of the project. Additional team reflections on both projects revealed benefit- and workload-dependent consideration of the feedback throughout the projects. There were also team members who never actively accessed the additional feedback or contributed to it due to a lack of personal incentives. Comparable usage has been observed during the design science research with students.

Answer to Research Question 2

The applied action research on two industrial software projects revealed that the managers of the agile development teams actively utilized the socio-technical feedback during sprint retrospectives as a complementary asset to existing developmental information for team discussions of variations in performance outcomes. Retrospective feedback on communication behavior, moods, and productivity was more often accessed than the predictive and exploratory sprint support available. The observed variations in the use of and contributions to the socio-technical feedback throughout the projects depended on the feedback culture of the teams. The data reveal benefit- and workload-dependent utilization, mainly when incentives were not present for individual team members. Therefore, the utilization findings should not be generalized to other teams or projects.

The third research question of this thesis concerned the practical utility of socio-technical feedback in projects. Both design science research and action research observations improved the understanding of the team dynamics in the student and industrial software projects. The observations indicated that objective and subjective adjustments had been made to the teams based on the computer-aided feedback. The design science research found that student teams that actively used retrospective and predictive feedback achieved lower estimation errors and steadier development performance in the projects, thereby leading to higher customer satisfaction with the sprints. The exploratory sprint dependencies revealed additional behavioral patterns across the various sprints. For example, six student teams significantly improved their development behavior and achieved high customer satisfaction, particularly for the last sprint in the project. In prior sprints, only project manager feedback was considered relevant, although the customers also provided input. The sprint performance variations in three agile teams from software projects at Arvato SCM Solutions were characterized by the exploratory simulation model with regards to motivational, mood, and productivity influences. The simulations increased the project members' awareness, extended their understanding of earlier firefighting situations, and helped to overcome productivity bottlenecks during sprints.

For the action research with two industrial projects, the quantitative and qualitative observations revealed role-dependent use of the individual retrospective, predictive, and exploratory feedback modules. Especially in the sprint retrospective, the systematic visualization of socio-technical information closed knowledge gaps, allowing management and teams to jointly uncover patterns in their communication structures as well as link mood extrema with team performance deviations during sprints. Substantial performance deviations were particularly relevant to understand due to the changes (both planned and unplanned) in the availability of the team members throughout the projects. The predictive feedback supported additional awareness of communication behaviors and mood trends in the following weeks based on an automated ML model that best interprets the socio-technical patterns of the prior weeks, particularly in the event of significant changes or events. However, both teams only sporadically accessed the predictive and exploratory feedback modules due to a lack of time and incentives.

Answer to Research Question 3

The observations of the student projects revealed the added value based on objective and perceived developmental performance improvements with the active use of the computer-aided sprint feedback. The additional team reflections in the two industrial projects affirmed a team-dependent utility of socio-technical feedback. The reflections promoted thereof awareness and understanding of past and future sprint performances in terms of the underlying human factors. The simplified feedback visualizations enabled a common understanding of behavioral changes during the sprint retrospectives without requiring social-psychological expertise. Nevertheless, the team reflections on the use of computer-aided feedback are based on individual perceptions and project-dependent influences. Therefore, these findings should not be generalized to other teams or projects.

The evaluation of ProDynamics was conducted in the industry and focused on the applicability of computer-aided feedback support in two agile software projects. The quantitative and qualitative results of the case study indicate the general applicability of ProDynamics in agile projects and the usefulness of computer-aided feedback for understanding team dynamics in sprints without prior expertise regarding the social-psychological aspects. In addition, the practitioner reflections underscored the perceived practical relevance of socio-technical feedback and the need to better understand human factors in the software development process. However, the current feedback applicability is subject to some limitations.

7.1   Limitations of the Present Research

The concepts presented in this work concern the systematic examination of the relevant socio-technical aspects of agile software projects and computer-aided feedback on team dynamics and behavioral changes during sprints. The current version of the ProDynamics plugin for Jira supports the automated (i.e., interval-based) collection of socio-technical data as well as the processing and provision of retrospective, predictive, and exploratory sprint feedback assets for projects. ProDynamics enables development teams to examine and understand performance deviations during sprints that correspond to changes in team communication behaviors, interaction patterns, moods, and developmental productivity. Supplementary sprint feedback from customers and project managers was also systematically collected to enrich the teams' awareness of external perceptions of their performance in a timely manner. The integrated anomaly detection feature sends email notifications to project managers in case of significant team behavior changes.

The evaluation of the ProDynamics plugin through the two software projects revealed some limitations concerning the sociological data collection and use of computer-aided sprint feedback. Computer-aided sprint feedback builds on the notion that team members contribute their sociological perceptions via the weekly self-assessment surveys throughout the project. A reliable characterization of all the socio-technical dependencies considered in this work would only be possible if the respective team members contributed data to be used to examine the emergence of problems. The evaluation in one of the two projects revealed that the highly negative mood behaviors of two team members were not covered in the computer-aided sprint feedback. These two individuals never completed the weekly self-assessment surveys. Consequently, the other members' otherwise positive mood obscured the team's actual mood situation (see Section 6.4.2.3).

The ProDynamics plugin should not be considered as the holy grail for understanding all socio-technical aspects of agile software development. Nevertheless, it adds value in terms of automated data collection, processing, and timely provision of feedback on team behavior in ongoing projects. In this context, the evaluation indicated another limitation concerning the perceived practical utility of the feedback support. The retrospective, predictive, and exploratory feedback modules enabled the two development teams to cognitively examine several socio-technical dependencies as well as development performances from past sprints and future predictions or simulations (see Section 6.4). The primary reason for the low usage of the predictive and exploratory feedback modules was identified through the team interviews. The teams expected feedback mechanisms for concrete recommendations for future actions or improvements rather than model-based predictions or simulations of possible performance trends.

7.2   Prospects for Future Research

Several options from the current feedback concept were uncovered during the evaluation of ProDynamics for the two industrial projects and the observations made during the technological feasibility studies with the student software projects. The following improvements present visions for future research regarding computer-aided feedback on team dynamics in agile software projects.

The self-assessment methods used to systematically record sociological data by the individual team members cause the feedback concept to strongly depend on individual schedules (i.e., a lack of time due to workload) and the consistent participation of all the team members throughout the project. Cross-system methods can support the capture of sociological aspects complementary to the self-assessment surveys applied here. Such methods would allow the systematic processing of data that are natively accessible (e.g., observations of communication behaviors based on the media channel used, sentiment analysis of messages for supplementary insights into team mood) [123, 124]. The retrospective interaction belongs to the first approach in this work, which enables teams to review their task-related interaction patterns during sprints based on the natively tracked ticket activities in Jira.

Additional team meeting characteristics represent another critical point of consideration. The current version of ProDynamics covers duration, frequency, and attendance in terms of the meeting behavior of sprint teams. However, it does not provide qualitative insight into the efficiency of meetings. An adaption or interface for an existing software (e.g., Act4Teams short meeting analyses [123]) could add valuable meeting aspects to the socio-technical data foundation and thus also find consideration in the computer-aided feedback. Besides, the evaluation in practice showed that not every feedback module received equal attention. Feedback, which focuses on project roles, could promote a reduction or amalgamation of the currently complex functional scopes to those with the respective highest utility (e.g., to fulfill role-related tasks better).

The current utility of predictive and exploratory feedback is another weak point in teams' understanding of socio-technical aspects in sprints. The systematic processing and analysis of socio-technical data revealed that behavioral changes occurred in the teams during the projects. The current feedback methods do not provide explicit recommendations for actions based on the observed behaviors. Such advice must be determined through individually derived and updated heuristics due to the social nature of each team and its members (e.g., with the help of reinforcement learning). Reinforcement learning allows the feedback system to track recommended actions regarding the perceived success/improvement or failure/drawback of an applied action, thus allowing a systematic adjustment of the recommendation heuristic over time.

Appendix A

Metrics and Issue Details in Agile Software Development

A.1 Common Metrics in Agile Software Development

A.2 Supplementary Issue Details Covered in Jira

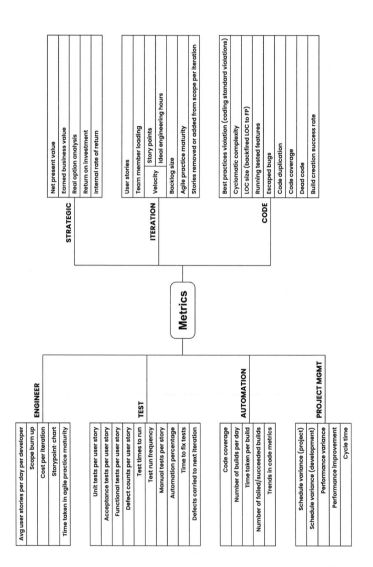

Table A.1: Set of Metrics in Agile Software Development [178]

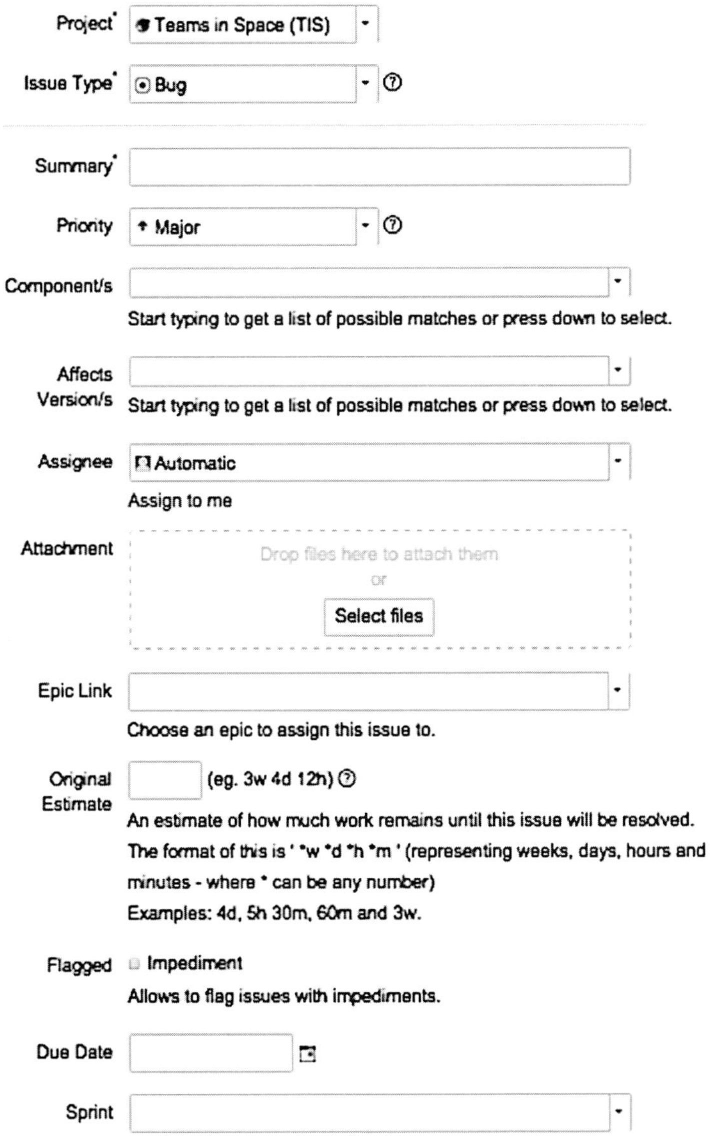

Figure A.2: Supplementary Issue Details Covered in Jira

Appendix B

System Dynamics Modeling Documents

B.1   Model Conceptualization Documents

B.2   Modular Structure of Sprint Dynamics Model

Interviewer:_____   ID:_____

FLOW Elicitation Fragebogen

Phase: Erheben Technik: Informationsflüsse elicitieren

Name	
Kontakt (✉ / ☎)	
Abteilung / Gruppe	
Datum	Start: Ende:

1. Allgemein

1. Wie lange sind Sie schon Teil Ihrer Abteilung / Gruppe?

2. Ausbildungshintergrund?

3. Professioneller Werdegang?

4. Ihr Arbeitsplatz? (Wie viele im Büro? Wie oft Störungen?)

5. Ihre Hauptaufgabe? Tätigkeitsbezeichnung?

6. Welche Tools, Prozesse, Programmiersprachen benutzen Sie während Ihrer täglichen Arbeit?

2. Informationsflüsse

Benutzen Sie die *Erhebungsbögen* für diesen Teil und fassen Sie unten zusammen. Zudem können die folgenden Fragen als Anleitung zum Ausfüllen des Erhebungsbogens genutzt werden.

2.1. *FLOW-Interface (Erhebungsbogen)*

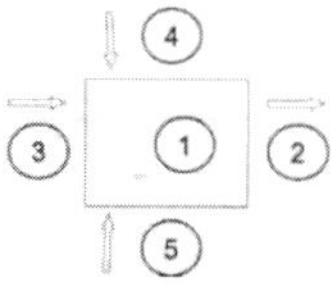

1. Was sind Ihre Hauptaufgaben? -> je ein Bogen

2. Wer nutzt Ihre Arbeitsergebnisse? In welcher Form/Medium? -> Output

3. Welche Informationen benötigen Sie für Ihre Arbeit? → Input

SE Fachgebiet Software Engineering Leibniz   1/2   02.10.2009
Universität Hannover DFG Projekt   Kurt Schneider & Kai Stapel
InfoFLOW, 2008 - 2011   www.se.uni-hannover.de/forschung/flow

Figure B1.1: FLOW Interview Template - Page 1

Interviewer:_____ ID:_____

4. Müssen Sie bei Ihrer Arbeit bestimmten Vorgaben folgen? (mündlich/schriftlich)
   -> Steuerung

5. Welche Werkzeuge oder Personen unterstützen die Durchführung Ihrer Arbeit?
   -> Unterstützung

2.2. *Flüssige Information und Erfahrung*
Weitere Fragen, um andere Informationsquellen und -senken aufzudecken:

1. Gibt es andere Stakeholder Ihrer Arbeit?

2. Mit wem arbeiten Sie zusammen?

3. Welche Meetings/Gespräche/Abstimmungen sind für die Funktionen wichtig?

4. Wer prüft und wogegen? Was geschieht bei Ablehnung?

5. Wie werden die Ergebnisse festgehalten, geprüft, freigegeben?

6. Wer hat auf dem Gebiet die meisten Erfahrungen? Wie profitieren Sie u. die anderen davon?

7. Wie viel Prozent der Informationen bekommen Sie schriftlich?

8. Wie viel Prozent der Informationen bekommen Sie mündlich?

9. Wie viele Informationen fließen über die Gruppenleiter?

3. Allgemein
1. Was funktioniert gut?

2. Wo gibt es Probleme?

Figure B1.2: FLOW Interview Template - Page 2

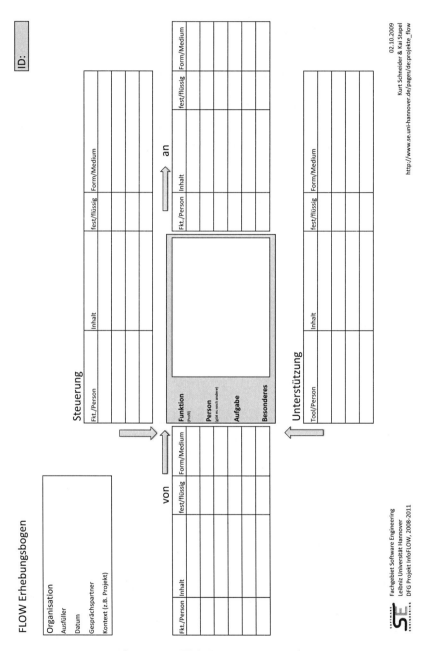

Figure B1.3: FLOW Interview Template

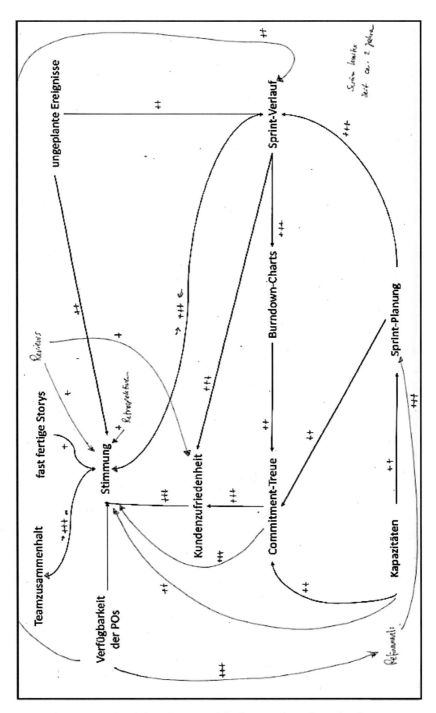

Figure B1.4: Workshop Excerpt of the Intermediate Causality Rating

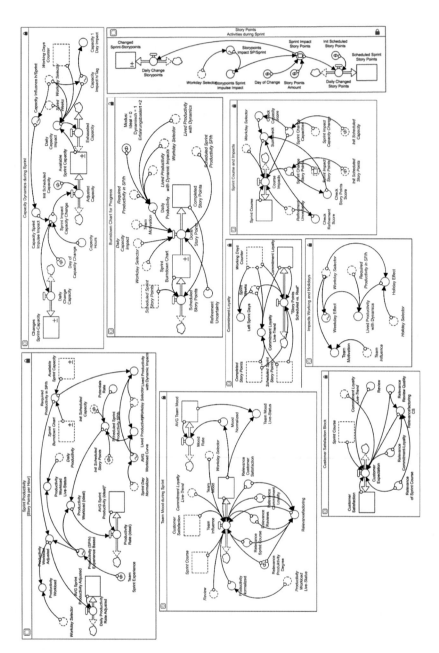

Table B2.1: Modular Structure of Sprint Dynamics Model

Appendix C

GQM-Related Documentation

C.1   Defined Goals (G1-G4)

C.2   Defined Abstraction Sheets (AS2-AS4)

C.3   Defined Operational Question Set

C.4   Defined Quantitative Metrics

C.5   Overview as GQM-Model

Table C1.1: Goal-Definition Template for G1

Analyze	Team Mood
For the Purpose of	an Improved Understanding
With respect to	Ppositive and Negative Moods
From the Viewpoint of	Development Teams
In the Context of	ASD

Table C1.2: Goal-Definition Template for G2

Analyze	Team Communication
For the Purpose of	an Improved Understanding
With respect to	Structures, Intensity and Media Channel
From the Viewpoint of	Development Teams
In the Context of	ASD

Table C1.3: Goal-Definition Template for G3

Analyze	Sprint Satisfaction
For the Purpose of	an Improved Understanding
With respect to	external Perceptions
From the Viewpoint of	Stakeholder and Scrum Master
In the Context of	ASD

Table C1.4: Goal-Definition Template for G4

Analyze	Team Performance
For the Purpose of	an Improved Understanding
With respect to	Progress, Workload and Problems
From the Viewpoint of	Development Teams
In the Context of	ASD

Object:	Purpose:	Quality focus:	Perspective:	Context:
Team Communication	Understanding	Communication Changes	Development Team	ASD-Process

Quality Focus:

Knowing the communication changes in teams requires understanding <u>organizational structures</u> and <u>interactions</u>, the <u>communication intensity</u> and used <u>media channel</u>.

Variation Factors:

Comparable <u>objective</u> and <u>subjective measures</u> about simple <u>communication characteristics</u> in teams.

Baseline Hypothesis:

Team communication changes over time are often <u>not</u> (consistently) tracked or systematically characterized.

Impact on Baseline Hypothesis:

The <u>shared understanding</u> in teams for communication changes involving <u>standardized measurements</u> (and network visualization) enables <u>determining (dys)functional structures</u> faster.

Feedback:
A steady team communication and structures are crucial for project and team success.

ID: AS2

Figure C2.1: Abstraction Sheet (AS2)

Object:	Purpose:	Quality focus:	Perspective:	Context:
Sprint Satisfaction	Understanding	Satisfaction Status	Stakeholder, Scrum Master	ASD-Process

Quality Focus:

Knowing the <u>satisfaction status</u> about <u>the performance</u> and <u>the product</u> requires understanding the perceptual differences from <u>external viewpoints</u> (e.g., the Stakeholder and Scrum Master).

Variation Factors:

<u>Systematic capture</u> of Sprint satisfaction feedback through a <u>standardized team performance</u> and <u>product rating scale</u>.

Baseline Hypothesis:

A comparison of external team feedback concerning the satisfaction of Stakeholder and Scrum Master in Sprints is often only <u>superficially managed without a comparable (quantitative)</u> data foundation.

Impact on Baseline Hypothesis:

<u>Early indication of false perceptions</u> concerning the Sprint success of teams based on comparable Stakeholder and Scrum Master ratings.

Feedback:
Agile teams depend on frequent internal and external feedback as an early indicator for Sprint success and improvements.

ID: AS3

Figure C2.2: Abstraction Sheet (AS3)

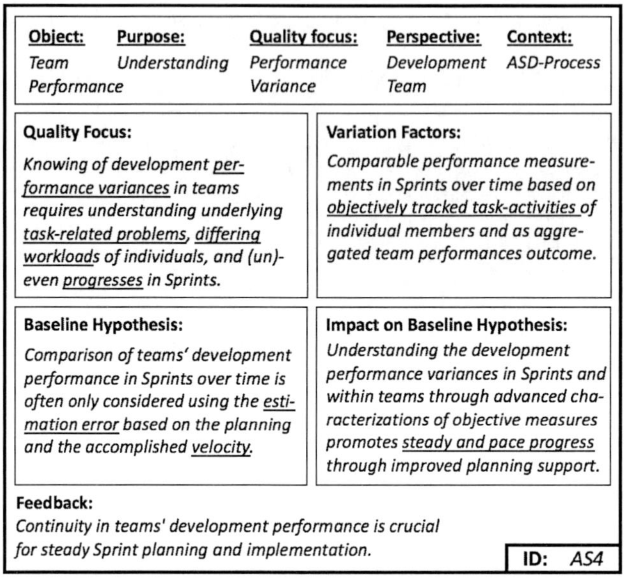

Object:	Purpose:	Quality focus:	Perspective:	Context:
Team Performance	Understanding	Performance Variance	Development Team	ASD-Process

Quality Focus:

Knowing of development performance variances in teams
requires understanding underlying
task-related problems, differing
workloads of individuals, and (un)-
even progresses in Sprints.

Variation Factors:

Comparable performance measurements in Sprints over time based on
objectively tracked task-activities of
individual members and as aggregated team performances outcome.

Baseline Hypothesis:

Comparison of teams' development
performance in Sprints over time is
often only considered using the estimation error based on the planning
and the accomplished velocity.

Impact on Baseline Hypothesis:

Understanding the development
performance variances in Sprints and
within teams through advanced characterizations of objective measures
promotes steady and pace progress
through improved planning support.

Feedback:
Continuity in teams' development performance is crucial
for steady Sprint planning and implementation.

ID: AS4

Figure C2.3: Abstraction Sheet (AS4)

Table C2.5: Defined Question Set for $G1$

Question	Goal-Question-ID
How intense is the perceived negative mood?	*G1.Q1*
How intense is the perceived positive mood?	*G1.Q2*
How is the level of mood in teams?	*G1.Q3*

Table C2.6: Defined Question Set for $G2$

Question	Goal-Question-ID
How is the communication structure in teams?	*G2.Q4*
How is the communication intensity in teams?	*G2.Q5*
What communication channel are used?	*G2.Q6*

Table C2.7: Defined Question Set for $G3$

Question	Goal-Question-ID
How do customer perceive the sprint performance?	*G3.Q7*
How do project leader perceive the sprint performance?	*G3Q8*

Table C2.8: Defined Question Set for $G4$

Question	Goal-Question-ID
How steady is the development performance?	*G4.Q9*
How balanced do teams develop?	*G4.Q10*
What is the planning error in sprints?	*G4.Q11*

(M1) Velocity Metric:

$$n = \{\text{Teamsize}\} \tag{C4.1}$$

$$j \in \{\text{Developer}_1, \ldots, \text{Developer}_n\} \tag{C4.2}$$

$$\text{velocity}_{developer}(j) = \frac{\text{Done [SP]}_j}{\text{Estimated [SP]}_j} \tag{C4.3}$$

$$\text{velocity}_{team} = \frac{\sum_{j=1}^{n} \text{velocity}_{developer}(j)}{n} \tag{C4.4}$$

(M2) Estimation Error:

$$\text{estimationError} = \left| 1 - \frac{\text{DDone [SP]}}{\text{[SP]-Scheduled} \pm \text{[SP]-Changes}} \right| \tag{C4.5}$$

(M3) Workload Balance:

$$n = \{\text{Teamsize}\} \tag{C4.6}$$

$$j \in \{\text{Developer}_1, \ldots, \text{Developer}_n\} \tag{C4.7}$$

$$\text{workload}_{\bar{x}} = \frac{\sum_{j=1}^{n} \text{[SP]-Done}_j}{n} \tag{C4.8}$$

$$\text{workload}_{\sigma} = \sqrt{\frac{\sum_{j=1}^{n} \left(\text{[SP]-Done}_j - \text{workload}_{\bar{x}} \right)^2}{n}} \tag{C4.9}$$

$$\text{workload}_{extrema} = \left| \mathcal{P}\{(\text{[SP]-Done}_j - \text{workload}_{\bar{x}}) > \text{workload}_{\sigma}\} \right| \tag{C4.10}$$

(M4) Sprint Satisfaction (Team Aspects):

$$\text{satRating}_i(j) \in \{1, \ldots, 5\}, \text{and } j \in \{\text{Customer, Scrum Master}\} \tag{C4.11}$$

$$i \in \{\text{informed, dedicated, organized, improving}\} \tag{C4.12}$$

$$\text{satTeam} = \text{med}_{i,j}(\text{satRating}_i(j)) \tag{C4.13}$$

(M5) Sprint Satisfaction (Product Aspects):

$$\text{satRating}_i(j) \in \{1, \ldots, 5\}, \text{and } j \in \{\text{Customer, Scrum Master}\} \tag{C4.14}$$

$$i \in \{\text{Goal Fulfilled, -Exceeded, Innovative, Claims}\} \tag{C4.15}$$

$$\text{satProduct} = \text{med}_{i,j}(\text{satRating}_i(j)) \tag{C4.16}$$

(M6) Positive Affect:

$$\text{moodRating}_i(j) \in \{1, \dots, 5\}, \tag{C4.17}$$

$$i \in \{\text{interested, awake, elated, happy, strong, active}\} \tag{C4.18}$$

$$j \in \{\text{Developer}_1, \dots, \text{Developer}_n\}, \text{where } n = \{\text{Teamsize}\} \tag{C4.19}$$

$$\text{affPositive} = \text{med}_{i,j}(\text{moodRating}_i(j)) \tag{C4.20}$$

(M7) Negative Affect Metric:

$$\text{moodRating}_i(j) \in \{1, \dots, 5\}, \tag{C4.21}$$

$$i \in \{\text{tempted, angry, nervous, angsty, worried, confused}\} \tag{C4.22}$$

$$j \in \{\text{Developer}_1, \dots, \text{Developer}_n\}, \text{where } n = \{\text{Teamsize}\} \tag{C4.23}$$

$$\text{affNegative} = \text{med}_{i,j}(\text{moodRating}_i(j)) \tag{C4.24}$$

(M8) Task-related Problems Metrics:

$$\text{taskProblems}_i = \{\text{textual description} < 150 \text{ chars}\}, \tag{C4.25}$$

$$i = \{x \in \mathbb{N}_0\} \tag{C4.26}$$

$$\tag{C4.27}$$

(M9) Relationship Conflicts Metric:

$$\text{relationshipConflicts}_i = \{\text{textual description} < 150 \text{ chars}\}, \tag{C4.28}$$

$$i = \{x \in \mathbb{N}_0\} \tag{C4.29}$$

$$\tag{C4.30}$$

(M10) Meeting Duration (Weekly):

$$\text{meetingDuration}_{week}(j) \in \{<1h, 1\text{-}2h, 3\text{-}4h, 5\text{-}6h, 7h<\}, \tag{C4.31}$$

$$week \in \{\mathbb{N}_0\}, n = \{\text{Teamsize}\} \tag{C4.32}$$

$$j \in \{\text{Developer}_1, \dots, \text{Developer}_n\} \tag{C4.33}$$

$$\text{meetingDurationWeekly} = \text{med}_{week,j}(\text{meetingDuration}_{week}(j)) \tag{C4.34}$$

(M11) Meeting Quantity (Weekly):

$$\text{meetingQty}_{week}(j) \in \{\text{never, 1-2x, 3-4x, 5-6x, daily}\}, \qquad \text{(C4.35)}$$

$$week \in \{\mathbb{N}_0\}, n = \{\text{Teamsize}\} \qquad \text{(C4.36)}$$

$$j \in \{\text{Developer}_1, \dots, \text{Developer}_n\} \qquad \text{(C4.37)}$$

$$\text{meetingQtyWeekly} = \text{med}_{week,j}(\text{meetingQty}_{week}(j)) \qquad \text{(C4.38)}$$

(M12) Meeting Participation (Weekly):

$$\text{meetingQty}_{week}(j) \in \{\text{never, 1-2x, 3-4x, 5-6x, daily}\} \qquad \text{(C4.39)}$$

$$week \in \{\mathbb{N}_0\}, n = \{\text{Teamsize}\} \qquad \text{(C4.40)}$$

$$j \in \{\text{Developer}_1, \dots, \text{Developer}_n\} \qquad \text{(C4.41)}$$

$$\text{meetingQtyWeekly}_{team} = \text{med}_{week,j}(\text{meetingQty}_{week}(j)) \qquad \text{(C4.42)}$$

$$\text{meetingParticipation}_{dev}(j) = \frac{\text{meetingQty}_{week}(j)}{meetingQtyWeekly} \qquad \text{(C4.43)}$$

$$\text{meetingParticipation}_{team} = 1 - \frac{\text{meetingQty}_{week}(j)}{meetingQtyWeekly} \qquad \text{(C4.44)}$$

(M13) Media Usage:

$$\text{mediaChannel} = \begin{pmatrix} F2F \\ Video \\ Phone \\ Chat \\ Jira \\ Email \\ SMS \end{pmatrix} \qquad \text{(C4.45)}$$

$$\text{mediaWeight} = \begin{pmatrix} 4 \\ 3 \\ 2 \\ 2 \\ 1 \\ 1 \\ 1 \end{pmatrix} \qquad \text{(C4.46)}$$

$$\text{mediaChannel}_{maxUsage} = \sum_n (mediaWeight) * \frac{n * (n-1)}{2} \qquad \text{(C4.47)}$$

$$\text{mediaChannel}_{actualUsage} = \sum_{j=1}^{n} (\sum_{x=j+1}^{n} usedMediaChanel(Dev_j, Dev_x)) \qquad \text{(C4.48)}$$

(M14) Communication Intensity:

$$\text{comIntensity}(i,j,m) \in \{0,\ldots,4\} \tag{C4.49}$$
$$m = mediaChannel \tag{C4.50}$$
$$i,j \in \{1,\ldots,n = \{\text{Teamsize}\} \tag{C4.51}$$
$$\text{comIntensity}_i = med_{j,m}\{comIntensity(i,j,m)\} \tag{C4.52}$$
$$\tag{C4.53}$$

(M15) FLOW distance: (FD)

$$c(i,j) = \prod_m comIntensity(i,j,m) \cdot mediaWeight(m) \tag{C4.54}$$
$$m = mediaChannel \tag{C4.55}$$
$$i,j \in \{1,\ldots,n = \{\text{Teamsize}\} \tag{C4.56}$$
$$c_{max} = max_{int} \cdot mediaWeight \tag{C4.57}$$
$$FD(i,j) = 1 - \frac{c_{ij}}{c_{max}} \tag{C4.58}$$

(M16) FLOW centralization:

$$centrality(j) = \sum_i FD(i,j) \tag{C4.59}$$
$$centrality_{max} = max_j centrality(j) \tag{C4.60}$$
$$centralization = \sum_j \frac{centrality_{max} - centrality(j)}{maximumCentrality} \tag{C4.61}$$

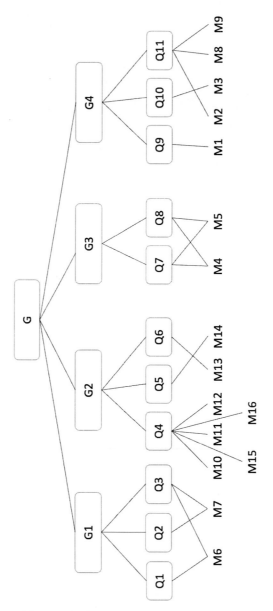

Figure C2.2: GQM-Model for the Socio-Technical Data Capture Concept

Appendix D

Sprint Feedback Support in Jira

D.1   Survey Manager

D.2   Sociological Team Survey

D.3   Customer/Project Manager Satisfaction Survey

D.4   Anomaly Notification

D.5   ProDynamics Feedback Assets

Prodynamics - Sprint Feedback

Managing Customer

Select Customer: ▾   Customer Summary

Add new Customer

Customer   **E-Mail**   **Project**

Fabian Kortum   @   INSTAGRAM1

Link a Customer with a Project   Overview

Customer Survey invitation/send

Select Project: EP-Instagram-1   ▾

Sprint	Status	
Exploration	Answered at 16.10.2020	
Iteration 1	Send on 2.11.2020	Submit
Iteration 2	Send on 7.0.2021	Submit
Polishing	Answered at 27.0.2021	

Summary of Projects/Latest Sprints

Managing Teamleader

Select Teamleader: ▾   Teamleader Summary

Add new Teamleader

Teamleader   **E-Mail**   **Project**

    @   INSTAGRAM1

Link Teamleader with a Project   Overview

Teamleader Survey invitation/send

Select Project: EP-Instagram-1   ▾   Reactivate E-Mail opt-in (Anomalies)

Sprint	Status	
Exploration	Answered at 16.10.2020	
Iteration 1	Send on 2.11.2020	Submit
Iteration 2	Send on 7.0.2021	Submit
Polishing	Answered at 27.0.2021	

Summary of Projects/Latest Sprints

Table D1.1: Survey Administration for External Contacts

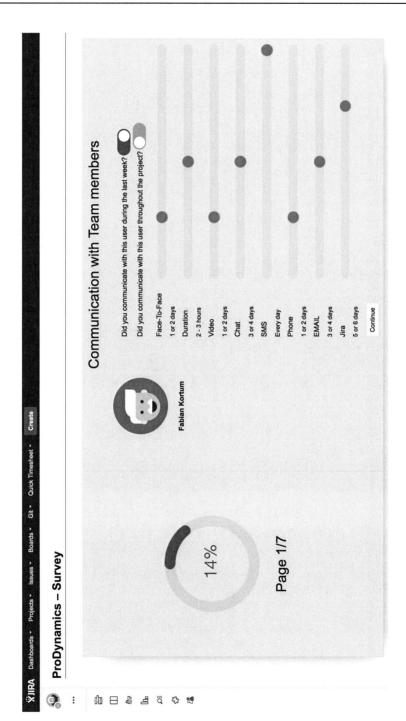

Table D2.1: Sociological Team Survey Excerpt Page 1

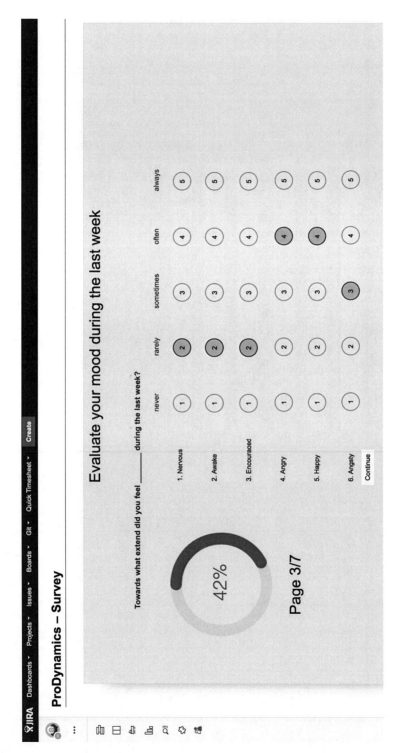

Table D2.2: Sociological Team Survey Excerpt Page 3

ProDynamics – Sprint Feedback

Survey for Customer (End of Sprint)

Sprint: Exploration Sprint Period: 19.10.2019 until 6.11.2019

Customer: Fabian Kortum   Project: Beispielprojekt_2019

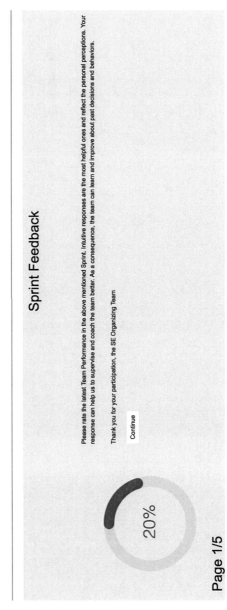

Sprint Feedback

Please rate the latest Team Performance in the above mentioned Sprint. Intuitive responses are the most helpful ones and reflect the personal perceptions. Your response can help us to supervise and coach the team better. As a consequence, the team can learn and improve about past decisions and behaviors.

Thank you for your participation, the SE Organizing Team

Continue

20%

Page 1/5

Table D3.1: Customers & Manager Satisfaction Survey in Jira Page 1

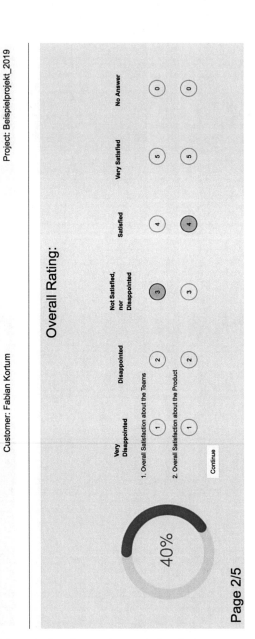

Table D3.2: Customers & Manager Satisfaction Survey in Jira Page 2

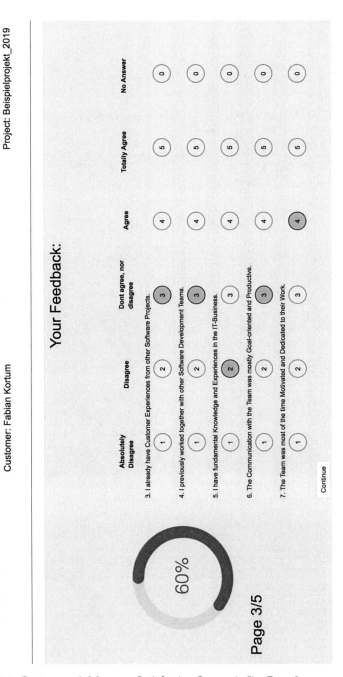

Table D3.3: Customers & Manager Satisfaction Survey in Jira Page 3

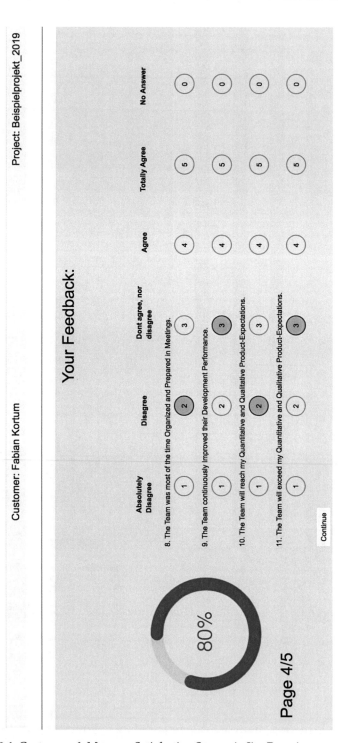

Table D3.4: Customers & Manager Satisfaction Survey in Jira Page 4

Survey for Customer (End of Sprint)

Sprint: Exploration Sprint Period: 19.10.2019 until 6.11.2019

Customer: Fabian Kortum

Project: Beispielprojekt_2019

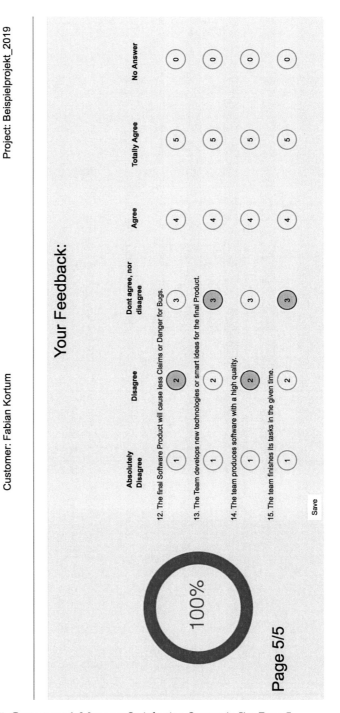

Table D3.5: Customers & Manager Satisfaction Survey in Jira Page 5

Dear **<name of project manager>**,

one or more anomalies in team behavior have been noticed during the last week for the project: **<name of the project>**

This is not necessarily negative, but as a project manager, it is advisable to evaluate the anomalies considering the following socio-technical aspects:

- **<Metric 1>** was **<lower>** than expected.
- **<Metric 2>** was **<higher>** than expected. (This is mostly an unwanted trend)

You can see the history of these metrics in the last few weeks in the following overview: **<link to view the temporal progress>**

-The Jira Administration-

If you do not want to receive anomaly notifications in the future, please click the following link: **<link to unsubscribe>**

Table D4.1: Example for an Automatic Anomaly Notification, cf. [169]

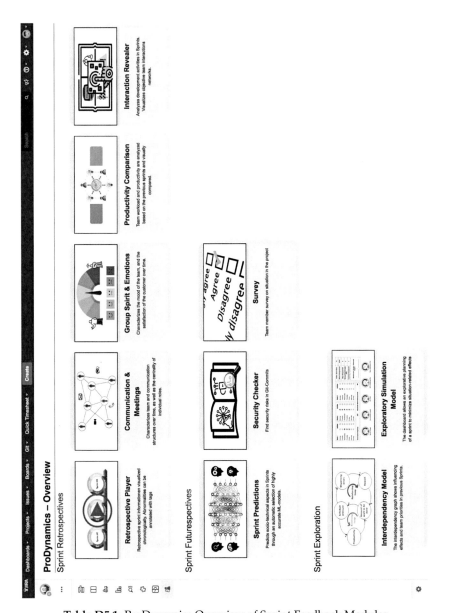

Table D5.1: ProDynamics Overview of Sprint Feedback Modules

Appendix E

Supervised Machine Learning

E.1 Model Settings

Table E1.1: Settings of Multiple Linear Regression Model

Description	Class for using linear regression for prediction.
Parameter	**Value/Setttings**
Attribute Selection Method	M5 method
Batch Size	100
Debug	False
Do Not Check Capabilities	False
Eliminate Colinear Atrributes	True
Minimal	False
Number of Decimal Places	4
Output Additional Stats	False
ridge	1.0E-08
Use QR Decomposition	False

Table E1.2: Settings of K Nearest Neighbor Model

Description	K nearest neighbors classifier with various distance measures applicable also to data with both numeric and nominal attributes.
Parameter	**Value/Settings**
Batch Size	100
Debug	False
Do Not Check Capabilities	False
Filter Neighbors Using Rules	False
Indexing	True
K	1
Learn Optimal K	True
Maximum K	100
Metric	City And Sample Value Difference
Number of Decimal Places	2
Vicinity Size For Density Based Metric	200
Voting	Inverse Square Distance
Weighting Method	Distance-Based

Table E1.3: Settings of Decision Tree Model

Description	Fast decision tree learner
Parameter	**Value Settings**
Batch Size	100
Debug	False
Do Not Check Capabilities	False
Initial Count	0.0
Minimum Number	2.0
minVarianceProp	0.001
No Pruning	False
Number of Decimal Places	2
Number of Folds	3
Seed	1
Spread Initial Count	False

Table E1.4: Setting of Support Vector Regression Model

Description	SMOreg implements the support vector machine for regression.
Parameter	**Value/Settings**
Batch Size	100
c	1.0
Debug	False
Do Not Check Capabilities	False
Filter Type	Normalize training data
Kernel	PolyKernel -E 1.0 -C 250007
Number of Decimal Places	2
Reg Optimizer	Reg SMO Improved -T 0.001 -V -P 1.0E-12 -L 0.001

Table E1.5: Settings of Multi-Layer Perceptron Model

Description	A classifier that uses backpropagation to learn a multi-layer perceptron to classify instances.
Parameter	**Value/Settings**
Auto Build	True
Batch Size	100
Debug	False
Decay	False
Do Not Check Capabilities	False
Hidden Layers	a
Learning Rate	0.3
Momentum	0.2
Nominal To Binary Filter	True
Normalize Attributes	True
Normalize Numeric Class	True
Number of Decimal Places	2
Reset	True
Resume	False
Seed	0
Training Time	500
Validation Set Size	0
Validation Threshold	20

Bibliography

[1] ABDEL-HAMID, T. K. ; MADNICK, S. E.: A Model of Software Project Management Dynamics. In: *COMPSAC '82 Proceedings*, 1982, S. 539–554

[2] ABDEL-HAMID, T. K. ; MADNICK, S. E.: *Software Project Dynamics: An Integrated Approach*. Prentice-Hall, Inc., 1991

[3] ABELEIN, U. ; PAECH, B. : State of Practice of User-Developer Communication in Large-Scale IT Projects. In: *International Working Conference on Requirements Engineering: Foundation for Software Quality*, 2014, S. 95–111

[4] ABILLA, P. : *Team Dynamics: Size Matters Redux*. 2006

[5] ABRAHAMSSON, P. ; SALO, O. ; RONKAINEN, J. ; WARSTA, J. : *Agile Software Development Methods: Review and Analysis*. VTT Technical Research Centre of Finland, 2002

[6] ACKOFF, R. L.: From Data to Wisdom. In: *Journal of Applied Systems Analysis* 16 (1989), Nr. 1, S. 3–9

[7] AGOSTINELLI, F. ; HOFFMAN, M. ; SADOWSKI, P. ; BALDI, P. : *Learning Activation Functions to Improve Deep Neural Networks*. arXiv:1412.6830, 2014

[8] ALBIN, S. ; FORRESTER, J. W. ; BREIEROVA, L. : *Building a System Dynamics Model: Part 1: Conceptualization*. MIT Press, 2001

[9] ALPAYDIN, E. : *Introduction to Machine Learning*. MIT Press, 2020

[10] AMORIM, L. F. ; MARINHO, M. ; SAMPAIO, S. : How (Un) happiness Impacts on Software Engineers in Agile Teams? In: *International Journal of Software Engineering & Applications (IJSEA)* 11 (2020), Nr. 3, S. 39–57

[11] AVISON, D. E. ; LAU, F. ; MYERS, M. D. ; NIELSEN, P. A.: Action Research. In: *Communications of the ACM* 42 (1999), Nr. 1, S. 94–97

[12] AWAD, M. ; KHANNA, R. : Support Vector Regression. In: *Efficient Learning Machines*. Springer, 2015, S. 67–80

[13] AYODELE, T. O.: Types of Machine Learning Algorithms. In: *New Advances in Machine Learning*. InTech, 2010, S. 19–48

[14] BALA, B. K. ; ARSHAD, F. M. ; NOH, K. M.: *System Dynamics*. Springer, 2017

[15] BARNARD, P. ; MAY, J. ; DUKE, D. ; DUCE, D. : Systems, Interactions, and Macrotheory. In: *ACM Transactions on Computer-Human Interaction (TOCHI)* 7 (2000), Nr. 2, S. 222–262

[16] BASILI, V. R. ; REITER JR, R. W.: An Investigation of Human Factors in Software Development. In: *Computer* 12 (1979), Nr. 12, S. 21–38

[17] BASILI VICTOR R, G. C. ; ROMBACH, H. D.: The Goal Question Metric Approach. In: *Encyclopedia of Software Engineering* (1994), S. 528–532

[18] BASKERVILLE, R. : Educing Theory from Practice. In: *Information Systems Action Research.* Springer, 2007, S. 313–326

[19] BASKERVILLE, R. L.: Investigating Information Systems with Action Research. In: *Communications of the Association for Information Systems* 2 (1999), Nr. 1, S. 19

[20] BATARSEH, F. A. ; GONZALEZ, A. J.: Predicting Failures in Agile Software Development through Data Analytics. In: *Software Quality Journal* 26 (2018), Nr. 1, S. 49–66

[21] BATTITI, R. : Using mutual information for selecting features in supervised neural net learning. In: *IEEE Transactions on neural networks* 5 (1994), Nr. 4, S. 537–550

[22] BAUM, T. ; KORTUM, F. ; SCHNEIDER, K. ; BRACK, A. ; SCHAUDER, J. : Comparing Pre-commit Reviews and Post-commit Reviews Using Process Simulation. In: *Journal of Software: Evolution and Process* 29 (2017), Nr. 11

[23] BECK, K. : *Extreme Programming Explained: Embrace Change.* Addison-Wesley, 2000

[24] BECK, K. ; BEEDLE, M. ; VAN BENNEKUM, A. ; COCKBURN, A. ; CUNNINGHAM, W. ; FOWLER, M. ; GRENNING, J. ; HIGHSMITH, J. ; HUNT, A. ; JEFFRIES, R. ; KERN, J. ; MARICK, B. ; MARTIN, R. C. ; MELLOR, S. ; SCHWABER, K. ; SUTHERLAND, J. ; THOMAS, D. : *Manifesto for Agile Software Development.* 2001

[25] BEECHAM, S. ; BADDOO, N. ; HALL, T. ; ROBINSON, H. ; SHARP, H. : Motivation in Software Engineering: A Systematic Literature Review. In: *Information and Software Technology* 50 (2008), Nr. 9, S. 860–878

[26] BÉLANGER, F. ; CROSSLER, R. E.: Privacy in the Digital Age: A Review of Information Privacy Research in Information Systems. In: *MIS Quarterly* 35 (2011), Nr. 4, S. 1017–1041

[27] BELLAZZI, R. ; ZUPAN, B. : Predictive Data Mining in Clinical Medicine: Current Issues and Guidelines. In: *International Journal of Medical Informatics* 77 (2008), Nr. 2, S. 81–97

[28] BEN-ZE'EV, A. : *The Subtlety of Emotions.* MIT Press, 2001

[29] BERENDT, B. ; GÜNTHER, O. ; SPIEKERMANN, S. : Privacy in E-commerce: Stated Preferences vs. Actual Behavior. In: *Communications of the ACM* 48 (2005), Nr. 4, S. 101–106

[30] BINDER, T. ; VOX, A. ; BELYAZID, S. ; HARALDSSON, H. ; SVENSSON, M. : Developing system dynamics models from causal loop diagrams. In: *Proceedings of the 22nd International Conference of the System Dynamic Society*, 2004, S. 1–21

[31] BISHOP, C. M.: *Pattern Recognition and Machine Learning*. Springer, 2006

[32] BJARNASON, E. : Distances between Requirements Engineering and Later Software Development Activities: A Systematic Map. In: *International Working Conference on Requirements Engineering: Foundation for Software Quality*, 2013, S. 292–307

[33] BJARNASON, E. ; WNUK, K. ; REGNELL, B. : Requirements Are Slipping through the Gaps — A Case Study on Causes & Effects of Communication Gaps in Large-Scale Software Development. In: *2011 IEEE 19th International Requirements Engineering Conference*, 2011, S. 37–46

[34] BLESS, M. : Distributed Meetings in Distributed Teams. In: *International Conference on Agile Software Development*, 2010, S. 251–260

[35] BOLL, M. L.: *Werkzeug zur Qualitätssicherung und grafischen Aufbereitung von empirischen Daten*. Leibniz Universität Hannover, Bachelorarbeit, 2017

[36] BOSTOCK, M. ; OGIEVETSKY, V. ; HEER, J. : D³ Data-Driven Documents. In: *IEEE Transactions on Visualization and Computer Graphics* 17 (2011), Nr. 12, S. 2301–2309

[37] BROOKS, F. P.: The Mythical Man-Month. In: *Datamation* 20 (1974), Nr. 12, S. 44–52

[38] BROWNLEE, J. : *Machine Learning Mastery with Weka*. 2018

[39] BUBSHAIT, A. A. ; FAROOQ, G. : Team Building and Project Success. In: *Cost Engineering* 41 (1999), Nr. 7, S. 34–38

[40] CANNON-BOWERS, J. A. ; SALAS, E. ; CONVERSE, S. : Shared Mental Models in Expert Team Decision Making. In: *Individual and Group Decision Making: Current Issues*. Lawrence Erlbaum Associates, Inc., 1993, S. 221–246

[41] CAO, L. ; RAMESH, B. ; ABDEL-HAMID, T. : Modeling Dynamics in Agile Software Development. In: *ACM Transactions on Management Information Systems* 1 (2010), Nr. 1, S. 5

[42] CARVER, J. ; JACCHERI, L. ; MORASCA, S. ; SHULL, F. : Issues in Using Students in Empirical Studies in Software Engineering Education. In: *Proceedings. 5th International Workshop on Enterprise Networking and Computing in Healthcare Industry*, 2003, S. 239–249

[43] CAWLEY, G. C. ; TALBOT, N. L.: Efficient Leave-One-Out Cross-Validation of Kernel Fisher Discriminant Classifiers. In: *Pattern Recognition* 36 (2003), Nr. 11, S. 2585–2592

[44] CHAU, T. ; MAURER, F. : Knowledge Sharing in Agile Software Teams. In: *Logic versus Approximation.* Springer, 2004, S. 173–183

[45] CHO, J. : Issues and Challenges of Agile Software Development with SCRUM. In: *Issues in Information Systems* 9 (2008), Nr. 2, S. 188–195

[46] CLARKE, R. : Internet Privacy Concerns Confirm the Case for Intervention. In: *Communications of the ACM* 42 (1999), Nr. 2, S. 60–67

[47] COCKBURN, A. ; HIGHSMITH, J. : Agile Software Development, the People Factor. In: *Computer* 34 (2001), Nr. 11, S. 131–133

[48] COCKBURN, A. : *Agile Software Development.* Addison-Wesley, 2002

[49] COCKBURN, A. : *Agile Software Development: The Cooperative Game.* Addison-Wesley, 2006

[50] COHN, M. : *Agile Estimating and Planning.* Pearson Education, 2005

[51] CULNAN, M. J. ; ARMSTRONG, P. K.: Information Privacy Concerns, Procedural Fairness, and Impersonal Trust: An Empirical Investigation. In: *Organization Science* 10 (1999), Nr. 1, S. 104–115

[52] CUMMINGS, J. N. ; CROSS, R. : Structural Properties of Work Groups and Their Consequences for Performance. In: *Social Networks* 25 (2003), Nr. 3, S. 197–210

[53] CURTIS, B. ; KRASNER, H. ; ISCOE, N. : A Field Study of the Software Design Process for Large Systems. In: *Communications of the ACM* 31 (1988), Nr. 11, S. 1268–1287

[54] DATTA, S. ; KAULGUD, V. ; SHARMA, V. S. ; KUMAR, N. : A Social Network Based Study of Software Team Dynamics. In: *Proceedings of the 3rd India Software Engineering Conference*, 2010, S. 33–42

[55] DAVIS, G. B. ; OLSON, M. H.: *Management Information Systems: Conceptual Foundations, Structure, and Development.* McGraw-Hill, Inc., 1984

[56] DE DREU, C. K. ; WEINGART, L. R.: Task versus Relationship Conflict, Team Performance, and Team Member Satisfaction: A Meta-Analysis. In: *Journal of Applied Psychology* 88 (2003), Nr. 4, S. 741–749

[57] DE'ATH, G. ; FABRICIUS, K. E.: Classification and Regression Trees: A Powerful Yet Simple Technique for Ecological Data Analysis. In: *Ecology* 81 (2000), Nr. 11, S. 3178–3192

[58] DELICADO, P. : Another Look at Principal Curves and Surfaces. In: *Journal of Multivariate Analysis* 72 (2001), Nr. 1, S. 84–116

[59] DERBY, E. ; LARSEN, D. ; SCHWABER, K. : *Agile Retrospectives: Making Good Teams Great.* Pragmatic Bookshelf, 2006

[60] DIENER, E. ; WIRTZ, D. ; TOV, W. ; KIM-PRIETO, C. ; CHOI, D.-w. ; OISHI, S. ; BISWAS-DIENER, R. : New Well-Being Measures: Short Scales to Assess Flourishing and Positive and Negative Feelings. In: *Social Indicators Research* 97 (2010), Nr. 2, S. 143–156

[61] DINGSØYR, T. ; HANSSEN, G. K.: Extending Agile Methods: Postmortem Reviews as Extended Feedback. In: *International Workshop on Learning Software Organizations*, 2002, S. 4–12

[62] DORAIRAJ, S. ; NOBLE, J. ; MALIK, P. : Understanding Team Dynamics in Distributed Agile Software Development. In: *International Conference on Agile Software Development*, 2012, S. 47–61

[63] DOWNEY, S. ; SUTHERLAND, J. : Scrum Metrics for Hyperproductive Teams: How They Fly like Fighter Aircraft. In: *2013 46th Hawaii International Conference on System Sciences*, 2013, S. 4870–4878

[64] DRUCKER, H. ; BURGES, C. J. ; KAUFMAN, L. ; SMOLA, A. J. ; VAPNIK, V. : Support Vector Regression Machines. In: *Advances in Neural Information Processing Systems* Bd. 9, 1997, S. 155–161

[65] DUMKE, R. (Hrsg.) ; ABRAN, A. (Hrsg.): *Software Measurement*. Deutscher Universitätsverlag, 1999

[66] DUNHAM, M. H.: *Data Mining: Introductory and Advanced Topics*. Pearson Education, 2006

[67] DYBÅ, T. ; DINGSØYR, T. : Empirical Studies of Agile Software Development: A Systematic Review. In: *Information and Software Technology* 50 (2008), Nr. 9–10, S. 833–859

[68] DYBÅ, T. ; PRIKLADNICKI, R. ; RÖNKKÖ, K. ; SEAMAN, C. ; SILLITO, J. : Qualitative Research in Software Engineering. In: *Empirical Software Engineering* 16 (2011), Nr. 4, S. 425–429

[69] ESPINOSA, J. A. ; SLAUGHTER, S. A. ; KRAUT, R. E. ; HERBSLEB, J. D.: Team Knowledge and Coordination in Geographically Distributed Software Development. In: *Journal of Management Information Systems* 24 (2007), Nr. 1, S. 135–169

[70] FAGERHOLM, F. ; IKONEN, M. ; KETTUNEN, P. ; MÜNCH, J. ; ROTO, V. ; ABRAHAMSSON, P. : Performance Alignment Work: How Software Developers Experience the Continuous Adaptation of Team Performance in Lean and Agile Environments. In: *Information and Software Technology* 64 (2015), S. 132–147

[71] FAYYAD, U. ; PIATETSKY-SHAPIRO, G. ; SMYTH, P. : From Data Mining to Knowledge Discovery in Databases. In: *AI Magazine* 17 (1996), Nr. 3, S. 37

[72] FERRARIO, M. A. ; SIMM, W. ; NEWMAN, P. ; FORSHAW, S. ; WHITTLE, J. : Software Engineering for 'Social Good': Integrating Action Research, Participatory Design, and Agile Development. In: *Companion Proceedings of the 36th International Conference on Software Engineering*, 2014, S. 520–523

[73] FEURER, M. ; KLEIN, A. ; EGGENSPERGER, K. ; SPRINGENBERG, J. ; BLUM, M. ; HUTTER, F. : Efficient and Robust Automated Machine Learning. In: CORTES, C. (Hrsg.) ; LAWRENCE, N. D. (Hrsg.) ; LEE, D. D. (Hrsg.) ; SUGIYAMA, M. (Hrsg.) ; GARNETT, R. (Hrsg.): *Advances in Neural Information Processing Systems* Bd. 28, 2015, S. 2962–2970

[74] FORRESTER, J. W.: Counterintuitive Behavior of Social Systems. In: *Theory and Decision* 2 (1971), Nr. 2, S. 109–140

[75] FORRESTER, J. W.: Lessons from System Dynamics Modeling. In: *System Dynamics Review* 3 (1987), Nr. 2, S. 136–149

[76] FORRESTER, J. W.: System Dynamics, Systems Thinking, and Soft OR. In: *System Dynamics Review* 10 (1994), Nr. 2–3, S. 245–256

[77] FOUNTAINE, A. ; SHARIF, B. : Emotional Awareness in Software Development: Theory and Measurement. In: *2017 IEEE/ACM 2nd International Workshop on Emotion Awareness in Software Engineering (SEmotion)*, 2017, S. 28–31

[78] FRANÇA, A. C. C. ; DA SILVA, F. Q. ; LC FELIX, A. de ; CARNEIRO, D. E.: Motivation in Software Engineering Industrial Practice: A Cross-Case Analysis of Two Software Organisations. In: *Information and Software Technology* 56 (2014), Nr. 1, S. 79–101

[79] FUKUNAGA, K. : *Introduction to Statistical Pattern Recognition*. Elsevier, 2013

[80] GARDNER, M. W. ; DORLING, S. : Artificial Neural Networks (The Multilayer Perceptron) — A Review of Applications in the Atmospheric Sciences. In: *Atmospheric Environment* 32 (1998), Nr. 14–15, S. 2627–2636

[81] GLAIEL, F. S. ; MOULTON, A. ; MADNICK, S. E.: Agile Project Dynamics: A System Dynamics Investigation of Agile Software Development Methods. In: *Proceedings of the 31st International Conference of the System Dynamics Society*, 2013, S. 1207

[82] GRAZIOTIN, D. ; FAGERHOLM, F. ; WANG, X. ; ABRAHAMSSON, P. : What Happens When Software Developers Are (Un) Happy. In: *Journal of Systems and Software* 140 (2018), S. 32–47

[83] GRAZIOTIN, D. ; WANG, X. ; ABRAHAMSSON, P. : Happy Software Developers Solve Problems Better: Psychological Measurements in Empirical Software Engineering. In: *PeerJ* 2 (2014), S. 289

[84] HALL, P. ; DEAN, J. ; KAYNAR-KABUL, I. ; SILVA, J. : An Overview of Machine Learning with SAS® Enterprise Miner™. In: *Proceedings of SAS Global Forum* (2014)

[85] HARTMANN, D. ; DYMOND, R. : Appropriate Agile Measurement: Using Metrics and Diagnostics to Deliver Business Value. In: *AGILE 2006*, 2006, S. 126–134

[86] HASSAN, A. E. ; XIE, T. : Software Intelligence: The Future of Mining Software Engineering Data. In: *Proceedings of the FSE/SDP Workshop on Future of Software Engineering Research*, 2010, S. 161–166

[87] HASTIE, T. ; TIBSHIRANI, R. ; FRIEDMAN, J. : *The Elements of Statistical Learning*. Springer, 2009

[88] HAUSKNECHT, J. P. ; HILLER, N. J. ; VANCE, R. J.: Work-Unit Absenteeism: Effects of Satisfaction, Commitment, Labor Market Conditions, and Time. In: *Academy of Management Journal* 51 (2008), Nr. 6, S. 1223–1245

[89] HENDERSON, J. C. ; LEE, S. : Managing I/S Design Teams: A Control Theories Perspective. In: *Management Science* 38 (1992), Nr. 6, S. 757–777

[90] HENG-LI, Y. ; JIH-HSIN, T. : Team Structure and Team Performance in Is Development: A Social Network Perspective. In: *Information & Management* 41 (2004), Nr. 3, S. 335–349

[91] HERBSLEB, J. D. ; MOCKUS, A. : An Empirical Study of Speed and Communication in Globally Distributed Software Development. In: *IEEE Transactions on Software Engineering* 29 (2003), Nr. 6, S. 481–494

[92] HEVNER, A. R.: A Three Cycle View of Design Science Research. In: *Scandinavian Journal of Information Systems* 19 (2007), Nr. 2, S. 4

[93] HEVNER, A. R. ; MARCH, S. T.: The Information Systems Research Cycle. In: *Computer* 36 (2003), Nr. 11, S. 111–113

[94] HEVNER, A. R. ; MARCH, S. T. ; PARK, J. ; RAM, S. : Design Science in Information Systems Research. In: *MIS Quarterly* 28 (2004), Nr. 1, S. 75–105

[95] HIGHSMITH, J. ; COCKBURN, A. : Agile Software Development: The Business of Innovation. In: *Computer* 34 (2001), Nr. 9, S. 120–127

[96] HINSZ, V. ; PARK, E. ; SJOMELING, M. : *Group Interaction Sustains Positive Moods and Diminishes Negative Moods*. Presented at the Annual Meeting of the Midwestern Psychological Association, Chicago, 2004

[97] HINTON, G. E. ; SEJNOWSKI, T. J.: *Unsupervised Learning: Foundations of Neural Computation*. MIT press, 1999

[98] HOEGL, M. ; GEMUENDEN, H. G.: Teamwork Quality and the Success of Innovative Projects: A Theoretical Concept and Empirical Evidence. In: *Organization Science* 12 (2001), Nr. 4, S. 435–449

[99] HOEGL, M. ; PARBOTEEAH, P. : Autonomy and Teamwork in Innovative Projects. In: *Human Resource Management: Published in Cooperation with the School of Business Administration, the University of Michigan and in Alliance with the Society of Human Resources Management* 45 (2006), Nr. 1, S. 67–79

[100] HOSSAIN, E. ; BABAR, M. A. ; PAIK, H. : Using Scrum in Global Software Development: A Systematic Literature Review. In: *2009 Fourth IEEE International Conference on Global Software Engineering*, 2009, S. 175–184

[101] HÖST, M. ; REGNELL, B. ; WOHLIN, C. : Using Students as Subjects – A Comparative Study of Students and Professionals in Lead-Time Impact Assessment. In: *Empirical Software Engineering* 5 (2000), Nr. 3, S. 201–214

[102] HOUGHTON, J. ; SIEGEL, M. ; GOLDSMITH, D. ; MOULTON, A. ; MADNICK, S. ; WIRSCH, A. : A Survey of Methods for Data Inclusion in System Dynamics Models: Methods, Tools and Applications. In: *Proceedings of the 32nd International Conference of the System Dynamics Society*, 2014, S. 1346

[103] HUMPHREY, W. S.: *Managing the Software Process*. Addison-Wesley, 1989

[104] HUO, M. ; VERNER, J. ; ZHU, L. ; BABAR, M. A.: Software Quality and Agile Methods. In: *Proceedings of the 28th Annual International Computer Software and Applications Conference* Bd. 1, 2004, S. 520–525

[105] HUTTER, F. ; KOTTHOFF, L. ; VANSCHOREN, J. : *Automated Machine Learning: Methods, Systems, Challenges*. Springer, 2019

[106] HYNDMAN, R. J. ; ATHANASOPOULOS, G. : *Forecasting: Principles and Practice*. OTexts, 2018

[107] IIVARI, J. ; VENABLE, J. R.: Action Research and Design Science Research-Seemingly Similar but Decisively Dissimilar. In: *ECIS 2009 Proceedings*, 2009, S. 73

[108] ILGEN, D. R. ; HOLLENBECK, J. R. ; JOHNSON, M. ; JUNDT, D. : Teams in Organizations: From Input-Process-Output Models to Imoi Models. In: *Annual Review of Psychology* 56 (2005), S. 517–543

[109] ISO/IEC: *ISO/IEC 25010:2011 Systems and Software Engineering – Systems and Software Quality Requirements and Evaluation (SQuaRE) – System and Software Quality Models*. International Organization for Standardization, 2011

[110] JERMAKOVICS, A. ; SILLITTI, A. ; SUCCI, G. : Mining and Visualizing Developer Networks from Version Control Systems. In: *Proceedings of the 4th International Workshop on Cooperative and Human Aspects of Software Engineering*, 2011, S. 24–31

[111] JOHN, M. ; MAURER, F. ; TESSEM, B. : Human and Social Factors of Software Engineering: Workshop Summary. In: *SIGSOFT Software Engineering Notes* 30 (2005), Nr. 4, S. 1–6

[112] JORDAN, M. I. ; MITCHELL, T. M.: Machine Learning: Trends, Perspectives, and Prospects. In: *Science* 349 (2015), Nr. 6245, S. 255–260

[113] JØRGENSEN, M. : Experience with the Accuracy of Software Maintenance Task Effort Prediction Models. In: *IEEE Transactions on Software Engineering* 21 (1995), Nr. 8, S. 674–681

[114] JØRGENSEN, M. ; GRUSCHKE, T. M.: The Impact of Lessons-Learned Sessions on Effort Estimation and Uncertainty Assessments. In: *IEEE Transactions on Software Engineering* 35 (2009), Nr. 3, S. 368–383

[115] KAUFFELD, S. ; LEHMANN-WILLENBROCK, N. : Meetings Matter: Effects of Team Meetings on Team and Organizational Success. In: *Small Group Research* 43 (2012), Nr. 2, S. 130–158

[116] KIM, M. ; ZIMMERMANN, T. ; DeLINE, R. ; BEGEL, A. : The Emerging Role of Data Scientists on Software Development Teams. In: *Proceedings of the 38th International Conference on Software Engineering*, 2016, S. 96–107

[117] KIRKMAN, B. L. ; ROSEN, B. : Beyond Self-Management: Antecedents and Consequences of Team Empowerment. In: *Academy of Management Journal* 42 (1999), Nr. 1, S. 58–74

[118] KIRKWOOD, C. W.: System Behavior and Causal Loop Diagrams. In: *System Dynamics Methods: A Quick Introduction*. Arizona State University, 1998, S. 1–14

[119] KLEIN, H. ; CANDITT, S. : Using Opinion Polls to Help Measure Business Impact in Agile Development. In: *Proceedings of the 1st International Workshop on Business Impact of Process Improvements*, 2008, S. 25–32

[120] KLIMOSKI, R. ; MOHAMMED, S. : Team Mental Model: Construct or Metaphor? In: *Journal of Management* 20 (1994), Nr. 2, S. 403–437

[121] KLONEK, F. ; GERPOTT, F. ; LEHMANN-WILLENBROCK, N. ; PARKER, S. : Time to Go Wild: How to Conceptualize and Measure Process Dynamics in Real Teams with High Resolution. In: *Organizational Psychology Review* 9 (2019), Nr. 4, S. 245–275

[122] KLONEK, F. ; MEINECKE, A. ; HAY, G. ; PARKER, S. : Capturing Team Dynamics in the Wild: The Communication Analysis Tool. In: *Small Group Research* 51 (2020), Nr. 3, S. 303–341

[123] KLÜNDER, J. : *Analyse der Zusammenarbeit in Softwareprojekten mittels Informationsflüssen und Interaktionen in Meetings*. Logos Verlag Berlin, 2019

[124] KLÜNDER, J. ; HORSTMANN, J. ; KARRAS, O. : Identifying the Mood of a Software Development Team by Analyzing Text-Based Communication in Chats with Machine Learning. In: *International Conference on Human-Centred Software Engineering* Springer, 2020, S. 133–151

[125] KLÜNDER, J. ; KARRAS, O. ; KORTUM, F. ; CASSELT, M. ; SCHNEIDER, K. : Different Views on Project Success. In: *International Conference on Product-Focused Software Process Improvement*, 2017, S. 497–507

[126] KLÜNDER, J. ; KARRAS, O. ; KORTUM, F. ; SCHNEIDER, K. : Forecasting Communication Behavior in Student Software Projects. In: *Proceedings of the The 12th International Conference on Predictive Models and Data Analytics in Software Engineering*, 2016, S. 1

[127] KLÜNDER, J. ; KARRAS, O. ; PRENNER, N. ; SCHNEIDER, K. : Modeling and Analyzing Information Flow in Development Teams as a Pipe System. In: *Third International Workshop on Human Factors in Modeling (HuFaMo 2018)*. CEUR-WS, 2018, S. 3–10

[128] KLÜNDER, J. ; KORTUM, F. ; ZIEHM, T. ; SCHNEIDER, K. : Helping Teams to Help Themselves: An Industrial Case Study on Interdependencies During Sprints. In: *Human-Centered Software Engineering*, 2019, S. 31–50

[129] KLÜNDER, J. ; PRENNER, N. ; WINDMANN, A.-K. ; STESS, M. ; NOLTING, M. ; KORTUM, F. ; HANDKE, L. ; SCHNEIDER, K. ; KAUFFELD, S. : Do You Just Discuss or Do You Solve? Meeting Analysis in a Software Project at Early Stages. In: *Proceedings of the 42nd International Conference on Software Engineering Workshops*, Association for Computing Machinery, 2020, S. 557–562

[130] KLÜNDER, J. ; SCHNEIDER, K. ; KORTUM, F. ; STRAUBE, J. ; HANDKE, L. ; KAUFFELD, S. : Communication in Teams - An Expression of Social Conflicts. In: *Human-Centered and Error-Resilient Systems Development: 6th International Conference on Human-Centered Software Engineering*, 2016, S. 111–129

[131] KOHAVI, R. : A Study of Cross-Validation and Bootstrap for Accuracy Estimation and Model Selection. In: *International Joint Conference on Artificial Intelligence* Bd. 14, 1995, S. 1137–1145

[132] KORKALA, M. ; ABRAHAMSSON, P. ; KYLLONEN, P. : A Case Study on the Impact of Customer Communication on Defects in Agile Software Development. In: *AGILE 2006*, 2006, S. 76–88

[133] KORTUM, F. : *An Experiences Survey about Sprint Feedback for Teams (1.0).* [Dataset] Zenodo, 2019

[134] KORTUM, F. ; KARRAS, O. ; KLÜNDER, J. ; SCHNEIDER, K. : Towards a Better Understanding of Team-Driven Dynamics in Agile Software Projects. In: *Product-Focused Software Process Improvement*, 2019, S. 725–740

[135] KORTUM, F. ; KLÜNDER, J. : Early Diagnostics on Team Communication: Experience-Based Forecasts on Student Software Projects. In: *10th International Conference on the Quality of Information and Communications Technology*, 2016, S. 166–171

[136] KORTUM, F. ; KLÜNDER, J. ; BRUNOTTE, W. ; SCHNEIDER, K. : Sprint Performance Forecasts in Agile Software Development - The Effect of Futurespectives on Team-Driven Dynamics. In: *31th International Conference on Software Engineering and Knowledge Engineering*, 2019, S. 94–128

[137] KORTUM, F. ; KLÜNDER, J. ; SCHNEIDER, K. : Don't Underestimate the Human Factors! Exploring Team Communication Effects. In: *International Conference on Product-Focused Software Process Improvement*, 2017, S. 457–469

[138] KORTUM, F. ; KLÜNDER, J. ; KARRAS, O. ; BRUNOTTE, W. ; SCHNEIDER, K. : Which Information Help Agile Teams the Most? An Experience Report on the Problems and Needs. In: *2020 46th Euromicro Conference on Software Engineering and Advanced Applications (SEAA)*, 2020, S. 306–313

[139] KORTUM, F. ; KLÜNDER, J. ; SCHNEIDER, K. : Behavior-Driven Dynamics in Agile Development: The Effect of Fast Feedback on Teams. In: *2019 IEEE/ACM International Conference on Software and System Processes (ICSSP)*, 2019, S. 34–43

[140] KRAIGER, K. ; WENZEL, L. H.: Conceptual Development and Empirical Evaluation of Measures of Shared Mental Models as Indicators of Team Effectiveness. In: *Team Performance Assessment and Measurement.* Psychology Press, 1997, S. 75–96

[141] KRASKOV, A. ; STÖGBAUER, H. ; GRASSBERGER, P. : Estimating Mutual Information. In: *Physical Review E* 69 (2004), Nr. 6

[142] KUHRMANN, M. (Hrsg.) ; TELL, P. (Hrsg.) ; KLÜNDER, J. (Hrsg.) ; HEBIG, R. (Hrsg.) ; LICORISH, S. (Hrsg.) ; MACDONELL, S. (Hrsg.): *HELENA Stage 2 Results*. [Online] ResearchGate, 2018

[143] KUPIAINEN, E. ; MÄNTYLÄ, M. V. ; ITKONEN, J. : Using Metrics in Agile and Lean Software Development — A Systematic Literature Review of Industrial Studies. In: *Information and Software Technology* 62 (2015), S. 143–163

[144] KWANTES, C. T. ; BOGLARSKY, C. A.: Perceptions of Organizational Culture, Leadership Effectiveness and Personal Effectiveness across Six Countries. In: *Journal of International Management* 13 (2007), Nr. 2, S. 204–230

[145] LALSING, V. ; KISHNAH, S. ; PUDARUTH, S. : People Factors in Agile Software Development and Project Management. In: *International Journal of Software Engineering & Applications* 3 (2012), Nr. 1, S. 117

[146] LAMOREUX, M. : Improving Agile Team Learning by Improving Team Reflections. In: *Agile Development Conference (ADC'05)*, 2005, S. 139–144

[147] LANDSBERGER, H. A.: *Hawthorne Revisited: Management and the Worker, Its Critics, and Developments in Human Relations in Industry*. Cornell University, 1958

[148] LAROSE, D. T.: *Discovering Knowledge in Data: An Introduction to Data Mining*. 2nd. Wiley Publishing, 2014

[149] LE, Q. P. Q.: *Tracen und Visualisieren von Entwickler – Interaktionen in Jira*. Leibniz Universität Hannover, Bachelorarbeit, 2019

[150] LEAU, Y. B. ; LOO, W. K. ; THAM, W. Y. ; TAN, S. F.: Software Development Life Cycle Agile vs Traditional Approaches. In: *International Conference on Information and Network Technology* Bd. 37, 2012, S. 162–167

[151] LEHMANN-WILLENBROCK, N. ; GROHMANN, A. ; KAUFFELD, S. : Task and Relationship Conflict at Work. In: *European Journal of Psychological Assessment* 27 (2011), Nr. 3, S. 171–178

[152] LEHMANN-WILLENBROCK, N. ; MEYERS, R. A. ; KAUFFELD, S. ; NEININGER, A. ; HENSCHEL, A. : Verbal Interaction Sequences and Group Mood: Exploring the Role of Team Planning Communication. In: *Small Group Research* 42 (2011), Nr. 6, S. 639–668

[153] LEHTONEN, T. ; ELORANTA, V.-P. ; LEPPANEN, M. ; ISOHANNI, E. : Visualizations as a Basis for Agile Software Process Improvement. In: *20th Asia-Pacific Software Engineering Conference (APSEC)* Bd. 1, 2013, S. 495–502

[154] LENBERG, P. ; FELDT, R. ; WALLGREN, L. G.: Human Factors Related Challenges in Software Engineering: An Industrial Perspective. In: *Proceedings of the Eighth International Workshop on Cooperative and Human Aspects of Software Engineering*, 2015, S. 43–49

[155] LI, P. : *Jira Essentials*. Packt Publishing, 2015

[156] LINDVALL, M. ; BASILI, V. R. ; BOEHM, B. ; COSTA, P. ; DANGLE, K. ; SHULL, F. ; TESORIERO, R. ; WILLIAMS, L. ; ZELKOWITZ, M. : Empirical Findings in Agile Methods. In: *Conference on extreme programming and agile methods*, 2002, S. 197–207

[157] LOH, W.-Y. : Classification and Regression Trees. In: *Wiley Interdisciplinary Reviews: Data Mining and Knowledge Discovery* 1 (2011), Nr. 1, S. 14–23

[158] MADACHY, R. J.: *Software Process Dynamics*. John Wiley & Sons, 2007

[159] MAHNIC, V. ; ZABKAR, N. : Using COBIT Indicators for Measuring Scrum-Based Software Development. In: *WSEAS Transactions on Computers* 7 (2008), Nr. 10, S. 1605–1617

[160] MALHOTRA, N. K. ; KIM, S. S. ; AGARWAL, J. : Internet Users' Information Privacy Concerns (IUIPC): The Construct, the Scale, and a Causal Model. In: *Information Systems Research* 15 (2004), Nr. 4, S. 336–355

[161] MARCH, S. T. ; SMITH, G. F.: Design and Natural Science Research on Information Technology. In: *Decision Support Systems* 15 (1995), Nr. 4, S. 251–266

[162] MARKUS, L. M.: Toward a Theory of Knowledge Reuse: Types of Knowledge Reuse Situations and Factors in Reuse Success. In: *Journal of Management Information Systems* 18 (2001), Nr. 1, S. 57–93

[163] MARSLAND, S. : *Machine Learning: An Algorithmic Perspective*. Chapman and Hall/CRC, 2011

[164] MARTIN, R. C.: *Agile Software Development: Principles, Patterns, and Practices*. Prentice Hall, 2002

[165] MCLEOD, R. ; SCHELL, G. P.: *Management Information Systems*. Pearson/Prentice Hall USA, 2007

[166] MELO OLIVEIRA, R. de ; GOLDMAN, A. : How to Build an Informative Workspace? An Experience Using Data Collection and Feedback. In: *2011 Agile Conference*, 2011, S. 143–146

[167] MENTZAS, G. ; APOSTOLOU, D. ; ABECKER, A. ; YOUNG, R. : *Knowledge Asset Management: Beyond the Process-Centred and Product-Centred Approaches*. Springer Science & Business Media, 2003

[168] MICHIE, D. (Hrsg.) ; SPIEGELHALTER, D. J. (Hrsg.) ; TAYLOR, C. (Hrsg.): *Machine Learning, Neural and Statistical Classification*. Paramount Publishing International, 1994

[169] MIRCEA, M. : *Konzeptionierung und Entwicklung eines JIRA-Plugins zur Erkennung von Anomalien im Teamverhalten während Sprints*. Leibniz Universität Hannover, Bachelorarbeit, 2021

[170] MOE, N. B. ; DINGSØYR, T. ; DYBÅ, T. : Understanding Self-Organizing Teams in Agile Software Development. In: *19th Australian Conference on Software Engineering (ASWEC 2008)*, 2008, S. 76–85

[171] MOE, N. B. ; DINGSØYR, T. ; DYBÅ, T. : Overcoming Barriers to Self-Management in Software Teams. In: *IEEE Software* 26 (2009), Nr. 6, S. 20–26

[172] MOE, N. ; DINGSØYR, T. : Scrum and Team Effectiveness: Theory and Practice. In: *Agile Processes in Software Engineering and Extreme Programming*, 2008, S. 11–20

[173] MOE, N. B. ; DINGSØYR, T. ; DYBÅ, T. : A Teamwork Model for Understanding an Agile Team: A Case Study of a Scrum Project. In: *Information and Software Technology* 52 (2010), Nr. 5, S. 480–491

[174] MORYS, P. : *Entwicklung eines Jira-Plugins: Digitale Erhebung und Nachverfolgung von Spint-Feedback zur Kundenzufriedenheit.* Leibniz Universität Hannover, Bachelorarbeit, 2019

[175] NAISBITT, J. : *Megatrends.* Warner Books, 1982

[176] NÄSLUND, D. : Logistics Needs Qualitative Research-Especially Action Research. In: *International Journal of Physical Distribution & Logistics Management* 32 (2002), Nr. 5, S. 321–338

[177] OSTERTAGOVÁ, E. : Modelling using Polynomial Regression. In: *Procedia Engineering* 48 (2012), S. 500–506

[178] OZA, N. ; KORKALA, M. : Lessons Learned in Implementing Agile Software Development Metrics. In: *UK Academy for Information Systems Conference Proceedings*, 2012, S. 38

[179] O'CONNOR, R. ; BASRI, S. : The Effect of Team Dynamics on Software Development Process Improvement. In: *International Journal of Human Capital and Information Technology Professionals (IJHCITP)* 3 (2012), Nr. 3, S. 13–26

[180] PAETSCH, P. ; EBERLEIN, A. ; MAURER, F. : Requirements Engineering and Agile Software Development. In: *WET ICE 2003. Proceedings. Twelfth IEEE International Workshops on Enabling Technologies: Infrastructure for Collaborative Enterprises*, 2003, S. 308–313

[181] PAL, S. K. ; MITRA, S. : Multilayer Perceptron, Fuzzy Sets, Classification. In: *IEEE Transactions on Neural Networks* 3 (1992), Nr. 5, S. 683–697

[182] PARSONS, H. M.: What Happened at Hawthorne?: New Evidence Suggests the Hawthorne Effect Resulted from Operant Reinforcement Contingencies. In: *Science* 183 (1974), Nr. 4128, S. 922–932

[183] PETERSEN, K. ; GENCEL, C. ; ASGHARI, N. ; BACA, D. ; BETZ, S. : Action Research as a Model for Industry-Academia Collaboration in the Software Engineering Context. In: *Proceedings of the 2014 International Workshop on Long-Term Industrial Collaboration on Software Engineering*, 2014, S. 55–62

[184] PETERSEN, K. ; WOHLIN, C.: A Comparison of Issues and Advantages in Agile and Incremental Development between State of the Art and an Industrial Case. In: *Journal of Systems and Software* 82 (2009), Nr. 9, S. 1479–1490

[185] PETERSEN, K. ; WOHLIN, C. :   Context in Industrial Software Engineering Research. In: *3rd International Symposium on Empirical Software Engineering and Measurement*, 2009, S. 401–404

[186] PFEFFER, J. : *Grundlagen der agilen Produktentwicklung: Basiswissen zu Scrum, Kanban, Lean Development*. BoD–Books on Demand, 2019

[187] PIATETSKY-SHAPIRO, G. ; BRACHMAN, R. J. ; KHABAZA, T. ; KLOESGEN, W. ; SIMOUDIS, E. : An Overview of Issues in Developing Industrial Data Mining and Knowledge Discovery Applications. In: *Conference on Knowledge Discovery and Data Mining* Bd. 96, 1996, S. 89–95

[188] PINTO, M. B. ; PINTO, J. K.:   Project Team Communication and Cross-Functional Cooperation in New Program Development. In: *Journal of Product Innovation Management: An International Publication of the Product Development & Management Association* 7 (1990), Nr. 3, S. 200–212

[189] POPPENDIECK, M. ; POPPENDIECK, T. : *Lean Software Development: An Agile Toolkit*. Addison-Wesley, 2003

[190] PROCACCINO, J. D. ; VERNER, J. M. ; SHELFER, K. M. ; GEFEN, D. : What Do Software Practitioners Really Think about Project Success: An Exploratory Study. In: *Journal of Systems and Software* 78 (2005), Nr. 2, S. 194–203

[191] PROVOST, F. ; FAWCETT, T. : Data Science and Its Relationship to Big Data and Data-Driven Decision Making. In: *Big Data* 1 (2013), Nr. 1, S. 51–59

[192] RANDERS, J. : *Elements of the System Dynamics Method*. Pegasus Communications, 1980

[193] RASCH, R. H. ; TOSI, H. L.:   Factors Affecting Software Developers' Performance: An Integrated Approach. In: *MIS Quarterly* 16 (1992), Nr. 3, S. 395–413

[194] RESHEF, D. N. ; RESHEF, Y. A. ; FINUCANE, H. K. ; GROSSMAN, S. R. ; MCVEAN, G. ; TURNBAUGH, P. J. ; LANDER, E. S. ; MITZENMACHER, M. ; SABETI, P. C.: Detecting Novel Associations in Large Data Sets. In: *Science* 334 (2011), Nr. 6062, S. 1518–1524

[195] RICHARDSON, G. P. ; PUGH, A. I.: *Introduction to System Dynamics Modeling with DYNAMO*. Productivity Press Inc., 1981

[196] ROBBINS, S. P. ; JUDGE, T. A.: *Organizational Behavior*. Pearson Education, 2013

[197] ROSENBLATT, F. : *Principles of Neurodynamics. Perceptrons and the Theory of Brain Mechanisms*. 1961

[198] RUIZ, M. ; SALANITRI, D. : Understanding How and When Human Factors Are Used in the Software Process: A Text-Mining Based Literature Review. In: *International Conference on Product-Focused Software Process Improvement*, 2019, S. 694–708

[199] RUMELHART, D. E. ; HINTON, G. E. ; WILLIAMS, R. J.: Learning Internal Representations by Error Propagation. In: *Parallel Distributed Processing: Explorations in the Microstructure of Cognition*. 1986, S. 318–362

[200] RUS, I. ; LINDVALL, M. ; SINHA, S. : Knowledge Management in Software Engineering. In: *IEEE Software* 19 (2002), Nr. 3, S. 26–38

[201] SALO, O. ; ABRAHAMSSON, P. : Empirical Evaluation of Agile Software Development: The Controlled Case Study Approach. In: *International Conference on Product Focused Software Process Improvement*, 2004, S. 408–423

[202] SANTOS, P. S. M. ; TRAVASSOS, G. H.: Action Research Can Swing the Balance in Experimental Software Engineering. In: *Advances in Computers* Bd. 83. 2011, S. 205–276

[203] SAWYER, S. ; ANNABI, H. : Methods as Theories: Evidence and Arguments for Theorizing on Software Development. In: *Social Inclusion: Societal and Organizational Implications for Information Systems*. 2006, S. 397–411

[204] SCHNEIDER, K. : *Experience and Knowledge Management in Software Engineering*. Springer Science & Business Media, 2009

[205] SCHNEIDER, K. ; KLÜNDER, J. ; KORTUM, F. ; HANDKE, L. ; STRAUBE, J. ; KAUFFELD, S. : Positive Affect through Interactions in Meetings: The Role of Proactive and Supportive Statements. In: *Journal of Systems and Software* 143 (2018), S. 59–70

[206] SCHNEIDER, K. ; LISKIN, O. : Exploring Flow Distance in Project Communication. In: *IEEE/ACM 8th International Workshop on Cooperative and Human Aspects of Software Engineering*, 2015, S. 117–118

[207] SCHNEIDER, K. ; LISKIN, O. ; PAULSEN, H. ; KAUFFELD, S. : Media, Mood, and Meetings: Related to Project Success? In: *ACM Transactions on Computing Education* 15 (2015), Nr. 4, S. 21

[208] SCHNEIDER, K. ; STAPEL, K. ; KNAUSS, E. : Beyond documents: visualizing informal communication. In: *Requirements Engineering Visualization* IEEE, 2008, S. 31–40

[209] SCHWABER, K. ; BEEDLE, M. : *Agile Software Development with Scrum*. Prentice Hall, 2002

[210] SEAMAN, C. B.: Qualitative Methods in Empirical Studies of Software Engineering. In: *IEEE Transactions on Software Engineering* 25 (1999), Nr. 4, S. 557–572

[211] SHARMA, S. : *Activation Functions in Neural Networks*. [Online] Towards Data Science, 2017

[212] SHARP, H. ; ROBINSON, H. : Some Social Factors of Software Engineering: The Maverick, Community and Technical Practices. In: *SIGSOFT Software Engineering Notes* 30 (2005), Nr. 4, S. 1–6

[213] SHARP, H. ; ROBINSON, H. : Collaboration and Co-ordination in Mature Extreme Programming Teams. In: *International Journal of Human-Computer Studies* 66 (2008), Nr. 7, S. 506–518

[214] SHORT, J. ; WILLIAMS, E. ; CHRISTIE, B. : *The Social Psychology of Telecommunications.* John Wiley & Sons, 1976

[215] SMOLA, A. J. ; SCHÖLKOPF, B. : A Tutorial on Support Vector Regression. In: *Statistics and Computing* 14 (2004), Nr. 3, S. 199–222

[216] STERMAN, J. : *Business Dynamics.* Irwin/McGraw-Hill, 2010

[217] STERMAN, J. D.: *System Dynamics Modeling for Project Management.* 1992

[218] STRAY, V. G. ; MOE, N. B. ; AURUM, A. : Investigating Daily Team Meetings in Agile Software Projects. In: *38th Euromicro Conference on Software Engineering and Advanced Applications,* 2012, S. 274–281

[219] SUDHAKAR, G. P. ; FAROOQ, A. ; PATNAIK, S. : Soft Factors Affecting the Performance of Software Development Teams. In: *Team Performance Management: An International Journal* 17 (2011), S. 187–205

[220] SUTHERLAND, J. ; SCHWABER, K. : *The Scrum Papers: Nuts, Bolts and Origins of an Agile Process.* 2007

[221] TAN, P.-N. ; STEINBACH, M. ; KUMAR, V. : *Introduction to Data Mining.* Pearson Education India, 2016

[222] TANG, J. ; DENG, C. ; HUANG, G.-B. : Extreme Learning Machine for Multilayer Perceptron. In: *IEEE Transactions on Neural Networks and Learning Systems* 27 (2015), Nr. 4, S. 809–821

[223] TAYLOR, R. : Interpretation of the Correlation Coefficient: A Basic Review. In: *Journal of Diagnostic Medical Sonography* 6 (1990), Nr. 1, S. 35–39

[224] TELLEGEN, A. ; WATSON, D. ; CLARK, L. A.: On the Dimensional and Hierarchical Structure of Affect. In: *Psychological Science* 10 (1999), Nr. 4, S. 297–303

[225] THOMPSON, E. R.: Development and Validation of an Internationally Reliable Short-Form of the Positive and Negative Affect Schedule (PANAS). In: *Journal of Cross-Cultural Psychology* 38 (2007), Nr. 2, S. 227–242

[226] THORNTON, C. ; HUTTER, F. ; HOOS, H. H. ; LEYTON-BROWN, K. : Auto-Weka: Combined Selection and Hyperparameter Optimization of Classification Algorithms. In: *Proceedings of the 19th ACM SIGKDD International Conference on Knowledge Discovery and Data Mining,* 2013, S. 847–855

[227] TIMOFEEV, R. : *Classification and Regression Trees (CART) Theory and Applications,* Humboldt University, Berlin, Diplomarbeit, 2004

[228] TORRECILLA-SALINAS, C. J. ; SEDEÑO, J. ; ESCALONA, M. ; MEJÍAS, M. : Estimating, Planning and Managing Agile Web Development Projects under a Value-Based Perspective. In: *Information and Software Technology* 61 (2015), S. 124–144

[229] TOUTENBURG, H. ; HEUMANN, C. : *Deskriptive Statistik: eine Einführung in Methoden und Anwendungen mit R und SPSS*. Springer-Verlag, 2008

[230] TREHUB, A. : *The Cognitive Brain*. MIT Press, 1991

[231] TRIPP, J. F. ; RIEMENSCHNEIDER, C. ; THATCHER, J. B.: Job Satisfaction in Agile Development Teams: Agile Development as Work Redesign. In: *Journal of the Association for Information Systems* 17 (2016), Nr. 4, S. 1

[232] VETRO, A. ; DÜRRE, R. ; CONOSCENTI, M. ; FERNÁNDEZ, D. M. ; JØRGENSEN, M. : Combining Data Analytics with Team Feedback to Improve the Estimation Process in Agile Software Development. In: *Foundations of Computing and Decision Sciences* 43 (2018), Nr. 4, S. 305–334

[233] VETRÒ, A. ; OGNAWALA, S. ; FERNÁNDEZ, D. M. ; WAGNER, S. : Fast Feedback Cycles in Empirical Software Engineering Research. In: *Proceedings of the 37th International Conference on Software Engineering* Bd. 2, 2015, S. 583–586

[234] WARE, C. : *Information Visualization: Perception for Design*. Elsevier, 2012

[235] WASSERMANN, S. ; FAUST, K. : *Social Network Analysis: Methods and Applications*. 1994

[236] WATSON, D. ; CLARK, L. A. ; TELLEGEN, A. : Development and Validation of Brief Measures of Positive and Negative Affect: The PANAS Scales. In: *Journal of Personality and Social Psychology* 54 (1988), Nr. 6, S. 1063

[237] WEN, J. ; LI, S. ; LIN, Z. ; HU, Y. ; HUANG, C. : Systematic Literature Review of Machine Learning Based Software Development Effort Estimation Models. In: *Information and Software Technology* 54 (2012), Nr. 1, S. 41–59

[238] WHITWORTH, E. ; BIDDLE, R. : The Social Nature of Agile Teams. In: *Agile 2007*, 2007, S. 26–36

[239] WILLIAMS, L. ; COCKBURN, A. : Agile Software Development: It's about Feedback and Change. In: *IEEE Computer* 36 (2003), Nr. 6, S. 39–43

[240] WITTEN, I. H. ; FRANK, E. ; HALL, M. A. ; PAL, C. J.: *Data Mining: Practical Machine Learning Tools and Techniques*. Morgan Kaufmann, 2016

[241] WOHLIN, C. ; RUNESON, P. ; HÖST, M. ; OHLSSON, M. C. ; REGNELL, B. ; WESSLÉN, A. : *Experimentation in Software Engineering*. Springer Science & Business Media, 2012

[242] WOLF, T. ; SCHROTER, A. ; DAMIAN, D. ; NGUYEN, T. : Predicting Build Failures Using Social Network Analysis on Developer Communication. In: *2009 IEEE 31st International Conference on Software Engineering*, 2009, S. 1–11

[243] WONG, T.-T. : Performance Evaluation of Classification Algorithms by K-Fold and Leave-One-Out Cross Validation. In: *Pattern Recognition* 48 (2015), Nr. 9, S. 2839–2846

[244] WROBEL, M. R.: Emotions in the Software Development Process. In: *6th International Conference on Human System Interactions*, 2013, S. 518–523

Curriculum Vitae

Fabian Kortum
born on March 31st, 1989 in Salzgitter, Germany

Professional Experience

since 08/2021	**Project Manager** for Vehicle Connectivity and Analytics IAV GmbH, Gifhorn
10/2015 – 03/2021	**Research Associate** at Software Engineering Group Leibniz University Hannover, Hannover
12/2013 – 10/2015	**Development Engineer** for Vehicle Systems and Analyses IAV GmbH, Gifhorn
10/2012 – 11/2013	**Student Associate** at the Institute for Computer Science AutoUni - Volkswagen AG, Wolfsburg
01/2009 – 09/2009	**Electronic Technician** for ECU-Quality Assurance Robert Bosch Elektronik GmbH, Salzgitter
09/2005 – 01/2009	**Electronics Technician Trainee** for Devices and Systems Robert Bosch Elektronik GmbH, Salzgitter

Education

10/2015 – 12/2021	**Doctor of Engineering** in Computer Science Leibniz University Hannover
09/2012 – 03/2015	**Master of Science** in Vehicle System Technologies University of Applied Sciences, Wolfenbüttel
09/2009 – 07/2012	**Bachelor of Science** in Technical Computer Science University of Applied Sciences, Wolfenbüttel and University of Wisconsin - Parkside, USA
09/2005 – 01/2009	**University of Applied Sciences Entrance Qualification** Technical Secondary School Fredenberg, Salzgitter

List of Scientific Publications

1. Kortum, F., Klünder, J., Karras, O., Brunotte, W., Schneider, K. : **Which Information Help Agile Teams the Most? An Experience Report on the Problems and Needs.** In: 46th Euromicro Conference on Software Engineering and Advanced Applications (SEAA), 2020, S. 306–313

2. Klünder, J., Prenner, N., Windmann, A.-K., Stess, M., Nolting, M. ; **Kortum, F.**, Handke, L., Schneider, K., Kauffeld, S. : **Do You Just Discuss or Do You Solve? Meeting Analysis in a Software Project at Early Stages.** In: Proceedings of the 42nd International Conference on Software Engineering Workshops, Association for Computing Machinery, 2020, S. 557–562

3. Kortum, F., Klünder, J., Brunotte, W., Schneider, K. : **Sprint Performance Forecasts in Agile Software Development - The Effect of Futurespectives on Team-Driven Dynamics.** In: 31th International Conference on Software Engineering and Knowledge Engineering, 2019, S. 94–128

4. Kortum, F., Karras, O., Klünder, J., Schneider, K. : **Towards a Better Understanding of Team-Driven Dynamics in Agile Software Projects.** In: Product-Focused Software Process Improvement, 2019, S. 725–740

5. Kortum, F., Klünder, J., Schneider, K. : **Behavior-Driven Dynamics in Agile Development: The Effect of Fast Feedback on Teams.** In: IEEE/ACM International Conference on Software and System Processes (ICSSP), 2019, S. 34–43

6. Kortum, F. : **An Experiences Survey about Sprint Feedback for Teams** (1.0). [Dataset] Zenodo, 2019

7. Klünder, J., **Kortum, F.**, Ziehm, T., Schneider, K. : **Helping Teams to Help Themselves: An Industrial Case Study on Interdependencies During Sprints.** In: Human-Centered Software Engineering, 2019, S. 31–50

8. Viertel, F., **Kortum, F.**, Wagner, L., Schneider, K. : **Are third-party libraries secure? A software library checker for java.** In: International Conference on Risks and Security of Internet and Systems, 2018, S. 18-34

9. Klünder, J., Karras, O., **Kortum, F.**, Casselt, M., Schneider, K. : **Different Views on Project Success.** In: International Conference on Product- Focused Software Process Improvement, 2017, S. 497–507

10. **Kortum, F.**, Klünder, J., Schneider, K. : **Don't Underestimate the Human Factors! Exploring Team Communication Effects.** In: International Conference on Product-Focused Software Process Improvement, 2017, S. 457–469

11. Baum, T., **Kortum, F.**, Schneider, K., Brack, A., Schauder, J. : **Comparing Pre-commit Reviews and Post-commit Reviews Using Process Simulation.** In: Journal of Software: Evolution and Process 29 (2017), Nr. 11

12. Klünder, J., Karras, O., **Kortum, F.**, Schneider, K. : **Forecasting Communication Behavior in Student Software Projects.** In: Proceedings of the The 12th International Conference on Predictive Models and Data Analytics in Software Engineering, 2016.

13. **Kortum, F.**, Klünder, J. : **Early Diagnostics on Team Communication: Experience-Based Forecasts on Student Software Projects.** In: 10th International Conference on the Quality of Information and Communications Technology, 2016, S. 166–171

14. Klünder, J., Schneider, K., **Kortum, F.**, Straube, J., Handke, L. ; Kauffeld, S. : **Communication in Teams - An Expression of Social Conflicts.** In: Human-Centered and Error-Resilient Systems Development: 6th International Conference on Human-Centered Software Engineering, 2016, S. 111–129